Number 301

George E. Andrews

Generalized Frobenius partitions

Memoirs

of the American Mathematical Society

Providence · Rhode Island · USA

May 1984 · Volume 49 · Number 301 (third of 5 numbers) · ISSN 0065-9266

Memoirs of the American Mathematical Society
Number 301

George E. Andrews

Generalized Frobenius partitions

Published by the
AMERICAN MATHEMATICAL SOCIETY
Providence, Rhode Island, USA

May 1984 · Volume 49 · Number 301 (third of 5 numbers)

MEMOIRS of the American Mathematical Society

This journal is designed particularly for long research papers (and groups of cognate papers) in pure and applied mathematics. It includes, in general, longer papers than those in the TRANSACTIONS.

Mathematical papers intended for publication in the Memoirs should be addressed to one of the editors. Subjects, and the editors associated with them, follow:

MEMOIRS are printed by photo-offset from camera-ready copy fully prepared by the authors. Prospective authors are encouraged to request booklet giving detailed instructions regarding reproduction copy. Write to Editorial Office, American Mathematical Society, P. O. Box 6248, Providence, Rhode Island 02940. For general instructions, see last page of Memoir.

SUBSCRIPTION INFORMATION. The 1984 subscription begins with Number 289 and consists of six mailings, each containing one or more numbers. Subscription prices for 1984 are $148 list; $74 member. A late charge of 10% of the subscription price will be imposed upon orders received from nonmembers after January 1 of the subscription year. Subscribers outside the United States and India must pay a postage surcharge of $10; subscribers in India must pay a postage surcharge of $15. Each number may be ordered separately; *please specify number* when ordering an individual number. For prices and titles of recently released numbers, refer to the New Publications sections of the NOTICES of the American Mathematical Society.

BACK NUMBER INFORMATION. For back issues see the AMS Catalogue of Publications.

TRANSACTIONS of the American Mathematical Society

This journal consists of shorter tracts which are of the same general character as the papers published in the MEMOIRS. The editorial committee is identical with that for the MEMOIRS so that papers intended for publication in this series should be addressed to one of the editors listed above.

Subscriptions and orders for publications of the American Mathematical Society should be addressed to American Mathematical Society, P. O. Box 1571, Annex Station, Providence, R. I. 02901. *All orders must be accompanied by payment.* Other correspondence should be addressed to P. O. Box 6248, Providence, R. I. 02940.

MEMOIRS of the American Mathematical Society (ISSN 0065-9266) is published bimonthly (each volume consisting usually of more than one number) by the American Mathematical Society at 201 Charles Street, Providence, Rhode Island 02904. Second Class postage paid at Providence, Rhode Island 02940. Postmaster: Send address changes to Memoirs of the American Mathematical Society, American Mathematical Society, P. O. Box 6248, Providence, RI 02940.

Table of Contents

		Page No.
1.	Introduction	1
2.	Frobenius's Idea	2
3.	The Jacobi Triple Product Identity	4
4.	Definitions and Simple Examples	6
5.	Generating Functions	9
6.	Finite Generating Functions	15
7.	Hecke Modular Forms	19
8.	q-Series	20
9.	Kloosterman Quadratic Forms	25
10.	Congruences	27
11.	The Dyson Rank for F-Partitions	31
12.	Partition Identities	33
13.	Asymptotics	35
14.	F. Franklin's Method	37
15.	Conclusion and Open Problems	39
16.	Tables	41
	References	42

Abstract

This paper is devoted to the study of equilength two-line arrays of non-negative integers. These are called generalized Frobenius partitions. It is shown that such objects have numerous interactions with modular forms, Kloosterman quadratic forms, the Lusztig-Macdonald-Wall conjectures as well as with classical theta functions and additive number theory.

1980 AMS Classification: 10A45

Key words: Partitions, modular forms.

Library of Congress Cataloging in Publication Data

Andrews, George E., 1938-
Generalized Frobenius partitions.

(Memoirs of the American Mathematical Society, ISSN 0065-9266 ; no. 301 (May 1984))
Includes bibliographical references.
1. Partitions (Mathematics) 2. Forms, Modular.
I. Title. II. Series: Memoirs of the American Mathematical Society ; no. 301.
QA3.A57 no. 301 [QA165] 510s [512'.72] 84-3059
ISBN 0-8218-2302-7

GENERALIZED FROBENIUS PARTITIONS

by

George E. Andrews(1)(2)

1. Introduction.

The object of this paper is to study two-rowed arrays of non-negative integers

$$(1.1) \qquad \begin{pmatrix} a_1 & a_2 & \cdots & a_r \\ b_1 & b_2 & \cdots & b_r \end{pmatrix}$$

wherein each row is of the same length and each is arranged in nonincreasing order. Such an object will be called a _generalized Frobenius partition_ or more simply an _F-partition_ of n if

$$(1.2) \qquad n = r + \sum_{i=1}^{r} a_i + \sum_{i=1}^{r} b_i .$$

Frobenius's name is invoked since he studied such objects [18; p. 523] in his work on group representation theory under the additional assumptions $a_1 > a_2 > \ldots > a_r \geqq 0$, $b_1 > b_2 > \ldots > b_r \geqq 0$.

This idea of Frobenius has been employed primarily in representation theory of the symmetric group (e.g. in [27], [32]). E. M. Wright [41], [42], [43] F. C. Auluck and H. N. V. Temperley (see [43] for complete references) have treated "stacks" which can be viewed as special cases of (1.1). Their

---

(1) Partially supported by National Science Foundation Grant MCS-8201733.

(2) The IBM symbolic manipulation language SCRATCHPAD (being field tested at Penn State) played a significant role in the discovery of many of the results in this paper.

Received by the editors August 17, 1982.

primary interest was the asymptotics of the enumerating functions.

Anytime one initiates an extensive study of new combinatorial objects such as F-partitions it is incumbent upon the author to provide rationale for their study. I was compelled to consider these objects by a study of the partition functions [6] associated with Regime III of Baxter's solution to the Hard Hexagon Model [10] (see related discussion in Section 12 below).

Having thus acknowledged the possible importance of F- partitions, we find on further examination that they are in a real sense the most natural combinatorial objects associated with the elliptic theta functions. This last assertion is perhaps best illustrated by the observation in Section 3 that Jacobi's Triple Product Identity [4; p. 21] becomes "obviouc" when viewed as an identity involving F-partitions; furthermore the point of view introduced in Section 3 provides the surprising generating function identities found in Sections 5 and 6. In addition the recent explication of Hecke modular forms via classical q-series results [7] has as its genesis the work in Section 7.

Next we should mention the Lusztig-Macdonald-Wall or L-M-W conjectures whose proof [5], [19] was so useful in Lusztig's pathbreaking work on group representation theory [29; p. 153]. In Section 8 we find that the L-M-W identities fit naturally into an infinite family of elegant identities related to F-partitions.

Finally as justification we note that many of the most appealing properties of the ordinary partition function $p(n)$ (and related restricted partition functions) carry over in quite nontrivial ways to F-partitions. For example, analogs of

$$p(5n + 4) \equiv 0 \pmod 5, \quad [31; p. 210],$$

are discussed in Section 10. The celebrated Dyson conjectures [16] (subsequently proved by Atkin and Swinnerton-Dyer [9]) appear to have elegant analogs for F-partitions which are very much open questions. The modular form nature of the many generating functions to be considered for various classes of F-partitions suggests that Hardy-Ramanujan-Rademacher asymptotic series exist for the related partition functions; this matter is briefly sketched in Section 13. Basil Gordon's beautiful generation of the Rogers-Ramanujan identities [20] translates naturally into a result about F-partitions. Section 14 illustrates how the combinatorial analysis of partitions into distinct parts given by Sylvester and his followers [17], [38] leads directly to corresponding results for F-partitions.

2. Frobenius's Idea.

When one considers partitions of an integer n in the classical manner, one starts with the idea of decomposing n into an unordered sum of positive integers. For example the seven partitions of 5 are $5, 4+1, 3+2, 3+1+1, 2+2+1, 2+1+1+1, 1+1+1+1+1$. Sylvester [38] attributed to N. M. Ferrers the notion of the <u>Ferrers graph</u> of a partition. For the partition $a_1 + a_2 + \dots + a_j$ (with $a_i \geq a_{i+1}$) this is generally represented as a set of left justified rows of equi-spaced dots wherein the i-th row has a_i dots. Thus the Ferrers graph of $7+7+5+4+2+2$ is

```
. . . . . . .
. . . . . . .
. . . . .
. . . .
. .
. .
```

The <u>conjugate</u> π^1 of a partition π is derived by counting dots in the columns of the Ferrers graph of π. Thus the conjugate of $7+7+5+4+2+2$ is $6+6+4+4+3+2+2$.

Frobenius desired a notation for partitions which would (among other things) exhibit immediately the conjugate of a partition π once π was given. To achieve this he considered the Ferrers graph of π. He deleted the main diagonal which possesses say r dots. The remaining rows of dots to the right of the diagonal are enumerated to provide one strictly decreasing sequence of r nonnegative integers (the r-th such row might be empty thus producing 0). The remaining dots below the diagonal are enumerated by columns to provide a second strictly decreasing sequence of r non-negative integers. The resulting two sequences are then presented in the Frobenius notation given by (1.1). For example the Frobenius notation for $7+7+5+4+2+2$ is $\begin{pmatrix} 6 & 5 & 2 & 0 \\ 5 & 4 & 1 & 0 \end{pmatrix}$ as is seen immediately from the Ferrers graph:

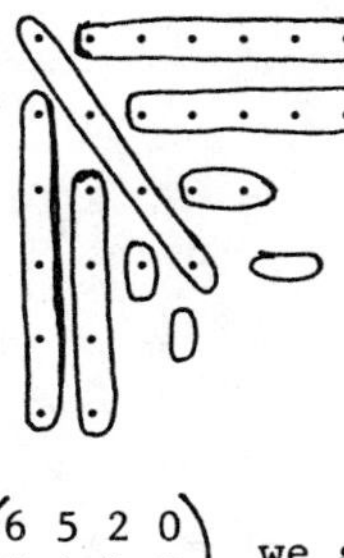

Now from the Frobenius symbol $\begin{pmatrix} 6 & 5 & 2 & 0 \\ 5 & 4 & 1 & 0 \end{pmatrix}$ we see immediately that the conjugate partition is $\begin{pmatrix} 5 & 4 & 1 & 0 \\ 6 & 5 & 2 & 0 \end{pmatrix}$. Also note that 27 is being partitioned and $27 = 4 + (6+5+2+0) + (5+4+1+0)$.

The notion that these "Frobenius symbols" are merely convenient guises of classical ordinary partitions helps to explain to some degree their being completely disregarded as objects in their own right. In the next section we shall see, in its simplest form, the device that opens the door to an extensive study of F-partitions.

3. The Jacobi Triple Product Identity and the General Principle.

We shall consider this identity in the following form:

$$(3.1) \qquad \prod_{n=1}^{\infty} (1 + zq^n)(1 + z^{-1}q^{n-1}) = \frac{1}{\prod_{n=1}^{\infty} (1-q^n)} \sum_{m=-\infty}^{\infty} z^m q^{\binom{m+1}{2}},$$

where $\binom{a}{2} = a(a-1)/2$, $z \neq 0$, $|q| < 1$.

There have been purely combinatorial proofs of (3.1) given by Sylvester [38], Cheema (see Gupta [23; pp. 213-215]), Wright [40], Sudler [37], Zolnowski [44] and Cohen [15]. These proofs deal with several different versions of (3.1), and indeed most of them can be described in the terms that Sudler [37; p. 67] uses to describe his two proofs, "... the two proofs given below are considerably more transparent than Sylvester's, though in a sense variants of it...".

It is our purpose now to prove (3.1) in a nearly transparent manner using F-partitions and a little (very little) analysis.

Let $\varphi(z)$ denote the left side of (3.1). Then

$$(3.2) \qquad \varphi(zq) = \prod_{n=1}^{\infty} (1+zq^{n+1})(1+z^{-1}q^{n-2}) = (1+z^{-1}q^{-1}) \prod_{n=1}^{\infty} (1+zq^{n+1})(1+z^{-1}q^{n-1})$$

$$= z^{-1}q^{-1} \prod_{n=1}^{\infty} (1+zq^n)(1+z^{-1}q^{n-1}) = z^{-1}q^{-1}\varphi(z).$$

Now $\varphi(z)$ can be expanded in a Laurent series in a deleted neighborhood of $z = 0$. Hence

$$(3.3) \qquad \varphi(z) = \sum_{n=-\infty}^{\infty} A_n(q)z^n = zq\varphi(zq) = \sum_{n=-\infty}^{\infty} A_n(q)z^{n+1}q^{n+1}.$$

Therefore for all integers n,

$$(3.4) \qquad A_n(q) = q^n A_{n-1}(q),$$

and so by appropriate iteration:

(3.5) $$A_n(q) = q^{\binom{n+1}{2}} A_0(q), \quad -\infty < n < \infty.$$

To complete the proof of (3.1) we need only show

(3.6) $$A_0(q) = \prod_{n=1}^{\infty} \frac{1}{1-q^n}.$$

However this is immediate using the correspondence described in Section 2 between ordinary partitions and F-partitions with strict decrease in rows. Namely it is well-known [4; p. 4] that

(3.7) $$\prod_{n=1}^{\infty} \frac{1}{1-q^n} = \sum_{n=0}^{\infty} p(n)q^n,$$

where $p(n)$ is the number of partitions of n.

How does the constant term in $\varphi(z)$ arise? Inspection of the left side of (3.1) shows that $q^N z^0$ arises, exactly as often as one finds pairs of terms $q^{a_1+a_2+\ldots+a_m} z^m$ with $a_1 > a_2 > \ldots > a_m \geqq 1$ arising from the first factor and $q^{b_1+b_2+\ldots b_m} z^{-m}$ arising from the second factor with $b_1 > b_2 > \ldots > b_m \geqq 0$. Wherein $(a_1+\ldots+a_m) + (b_1+\ldots+b_m) = N$. In other words, each contribution to q^N in the constant term arises from each F-partition of N of the form

$$\begin{pmatrix} (a_1-1) & (a_2-1) & \ldots & (a_m-1) \\ b_1 & b_2 & \ldots & b_m \end{pmatrix}$$

wherein $N = m + \sum_{i=1}^{m} (a_i-1) + \sum_{i=1}^{m} b_i$, with $a_1 - 1 > a_2 - 1 > \ldots > a_m - 1 \geqq 0$, $b_1 > b_2 > \ldots > b_m \geqq 0$. Since there is a bijection between F-partitions of this sort and ordinary partitions we see that (3.6) is established and consequently (3.1) is also.

It should be emphasized that this proof is also, in a sense, a variant of Sylvester's original proof. It is perhaps most closely related to E. M. Wright's proof although his change of variable in (3.1) obscures this fact.

The importance of the just completed proof is certainly not that yet another variation of the combinatorics of (3.1) has been found. The importance lies in the following observation:

* * *

GENERAL PRINCIPLE:

If $f_A(z,q) = f_A(z) = \Sigma P_A(m,n) z^m q^n$ denotes the generating function for $P_A(m,n)$ the number of ordinary partitions of n into m parts subject to the set of restrictions A, then

$$f_A(zq)f_B(z^{-1})$$

has as its constant term (i.e. coefficient of z^0) the generating function

$$\Phi_{A,B}(q) = \sum_{n\geq 0} \phi_{A,B}(n)q^n, \tag{3.8}$$

where $\varphi_{A,B}(n)$ is the number of F-partitions

$$\begin{pmatrix} a_1 & a_2 & \cdots & a_r \\ b_1 & b_2 & \cdots & b_r \end{pmatrix}$$

in which the top row is subject to the set of restrictions A and the bottom row is subject to the set of restrictions B.

* * *

If we denote by D the set of restrictions "distinct non-negative parts". Then the observation used above to prove Jacobi's identity is merely that

$$\phi_{D,D}(n) = p(n).$$

This simple observation embodied in our general principle allows us to study many general classes of F-partitions.

4. Definitions and Simple Examples.

We shall restrict our attention until Section 14 to two general classes of Frobenius partitions.

First we consider F-partitions that allow up to k repetitions of an integer in any row. Thus if A_k denotes the condition "each part repeated at most k times", then we shall treat

$$\Phi_k(q) \equiv \Phi_{A_k,A_k}(q) = \sum_{n\geq 0} \phi_{A_k,A_k}(n)q^n = \sum_{n\geq 0} \phi_k(n)q^n, \tag{4.1}$$

where we have for convenience replaced the subscript notation A_k,A_k by merely k.

We note $\phi_1(n) = p(n)$, the observation at the end of Section 3. The F-partitions enumerated by $\phi_2(3)$ are

$$(4.2)\qquad \begin{pmatrix}2\\0\end{pmatrix}\begin{pmatrix}0\\2\end{pmatrix}\begin{pmatrix}1\\1\end{pmatrix}\begin{pmatrix}10\\00\end{pmatrix}\begin{pmatrix}00\\10\end{pmatrix},$$

and

$$(4.3)\qquad \Phi_2(q) = 1 + q + 3q^2 + 5q^3 + 9q^4 + 14q^5 + 24q^6 + 35q^7 + 55q^8 + \cdots$$

The partitions enumerated by $\phi_3(4)$ are

$$(4.4)\qquad \begin{pmatrix}3\\0\end{pmatrix}\begin{pmatrix}0\\3\end{pmatrix}\begin{pmatrix}2\\1\end{pmatrix}\begin{pmatrix}1\\2\end{pmatrix}\begin{pmatrix}20\\00\end{pmatrix}\begin{pmatrix}00\\20\end{pmatrix}\begin{pmatrix}11\\00\end{pmatrix}\begin{pmatrix}00\\11\end{pmatrix}\begin{pmatrix}10\\10\end{pmatrix}\begin{pmatrix}100\\000\end{pmatrix}\begin{pmatrix}000\\100\end{pmatrix},$$

and

$$(4.5)\qquad \Phi_3(q) = 1 + q + 3q^2 + 6q^3 + 11q^4 + 18q^5 + 31q^6 + 49q^7 + 78q^8 + \cdots .$$

In addition

$$(4.6)\qquad \Phi_4(q) = 1 + q + 3q^2 + 6q^3 + 12q^4 + 20q^5 + 35q^6 + 56q^7 + 92q^8 + \cdots ,$$

and

$$(4.7)\qquad \Phi_5(q) = 1 + q + 3q^2 + 6q^3 + 12q^4 + 21q^5 + 37q^6 + 60q^7 + 99q^8 + \cdots .$$

The second general class of F-partitions at first glance seems rather contrived; however the resulting generating functions are at least as elegant in their properties as the $\Phi_k(q)$.

For this latter class we consider now k-copies of the nonnegative integers written "j_i" where $j \geqq 0$ and $1 \leqq i \leqq k$. We say $j_i < \ell_h$ precisely when $j < \ell$ or $j = \ell$ and $i < h$, and we say "j_i is distinct from ℓ_h" unless $j = \ell$ and $i = h$. Our condition A_k is now "each part distinct and taken from k copies of the integers." Thus

$$(4.8)\qquad C\Phi_k(q) \equiv \Phi_{A_k,A_k}(q) = \sum_{n\geqq 0} \Phi_{A_k,A_k}(n)q^n \quad \sum_{n\geqq 0} c\phi_k(n)q^n.$$

Sometimes the subscripts are thought of as providing different colors of each integer. Thus $c\phi_k(n)$ is called the number of F-partitions of n with k colors.

Of course one could generalize things still further by considering F-partitions with k colors and at most h repetitions in each row. This would provide a common generalization of (4.1) and (4.8); however it appears that the individual cases of most interest arise for $k = 1$ (e.g. (4.1)) or $h = 1$ (e.g. (4.8)).

The F-partitions enumerated by $c\phi_2(2)$ are

$$(4.9) \quad \begin{pmatrix} 1_1 \\ 0_1 \end{pmatrix}\begin{pmatrix} 1_2 \\ 0_1 \end{pmatrix}\begin{pmatrix} 1_1 \\ 0_2 \end{pmatrix}\begin{pmatrix} 1_2 \\ 0_2 \end{pmatrix}\begin{pmatrix} 0_1 \\ 1_1 \end{pmatrix}\begin{pmatrix} 0_1 \\ 1_2 \end{pmatrix}\begin{pmatrix} 0_2 \\ 1_1 \end{pmatrix}\begin{pmatrix} 0_2 \\ 1_2 \end{pmatrix}\begin{pmatrix} 0_2 & 0_1 \\ 0_2 & 0_1 \end{pmatrix},$$

and

$$(4.10) \quad C\Phi_2(q) = 1 + 4q + 9q^2 + 20q^3 + 42q^4 + 80q^5 + 147q^6 + 260q^7 + 445q^8 + \ldots$$

The partitions enumerated by $c\phi_3(2)$ are

$$\begin{pmatrix} 1_1 \\ 0_1 \end{pmatrix}\begin{pmatrix} 1_2 \\ 0_1 \end{pmatrix}\begin{pmatrix} 1_3 \\ 0_1 \end{pmatrix}\begin{pmatrix} 1_1 \\ 0_2 \end{pmatrix}\begin{pmatrix} 1_2 \\ 0_2 \end{pmatrix}\begin{pmatrix} 1_3 \\ 0_2 \end{pmatrix}\begin{pmatrix} 1_1 \\ 0_3 \end{pmatrix}\begin{pmatrix} 1_2 \\ 0_3 \end{pmatrix}\begin{pmatrix} 1_3 \\ 0_3 \end{pmatrix}$$

$$\begin{pmatrix} 0_1 \\ 1_1 \end{pmatrix}\begin{pmatrix} 0_1 \\ 1_2 \end{pmatrix}\begin{pmatrix} 0_1 \\ 1_3 \end{pmatrix}\begin{pmatrix} 0_2 \\ 1_1 \end{pmatrix}\begin{pmatrix} 0_2 \\ 1_2 \end{pmatrix}\begin{pmatrix} 0_2 \\ 1_3 \end{pmatrix}\begin{pmatrix} 0_3 \\ 1_1 \end{pmatrix}\begin{pmatrix} 0_3 \\ 1_2 \end{pmatrix}\begin{pmatrix} 0_3 \\ 1_3 \end{pmatrix}$$

$$\begin{pmatrix} 0_2 & 0_1 \\ 0_2 & 0_1 \end{pmatrix}\begin{pmatrix} 0_3 & 0_1 \\ 0_2 & 0_1 \end{pmatrix}\begin{pmatrix} 0_3 & 0_2 \\ 0_2 & 0_1 \end{pmatrix}\begin{pmatrix} 0_2 & 0_1 \\ 0_3 & 0_1 \end{pmatrix}\begin{pmatrix} 0_3 & 0_1 \\ 0_3 & 0_1 \end{pmatrix}\begin{pmatrix} 0_3 & 0_2 \\ 0_3 & 0_1 \end{pmatrix}$$

$$\begin{pmatrix} 0_2 & 0_1 \\ 0_3 & 0_2 \end{pmatrix}\begin{pmatrix} 0_3 & 0_1 \\ 0_3 & 0_2 \end{pmatrix}\begin{pmatrix} 0_3 & 0_2 \\ 0_3 & 0_2 \end{pmatrix},$$

and

$$(4.11) \quad C\Phi_3(q) = 1 + 9q + 27q^2 + 82q^3 + 207q^4 + 486q^5 + 1055q^6 + 2205q^7 + \ldots$$

In addition

$$C\Phi_4(q) = 1 + 16q + 68q^2 + 256q^3 + 777q^4 + 2160q^5 + 5460q^6 + 1305q^7 + \ldots$$

and

$$C\Phi_5(q) = 1 + 25q + 150q^2 + 675q^3 + 2450q^4 + 7876q^5 + 2285q^6 + 61550q^7 + \ldots$$

As is easily seen the coefficients in $C\Phi_k(q)$ are much larger than those in $\Phi_k(q)$. In the next section we shall proceed to find representations for both families of functions in terms of modular forms.

5. Generating Functions.

We begin with two central theorems that tie $\Phi_k(q)$ and $C\Phi_k(q)$ with multidimensional theta functions. We then find as corollaries that $\Phi_1(q)$, $\Phi_2(q)$, $\Phi_3(q)$, $C\Phi_1(q)$ and $C\Phi_2(q)$ are all representable by elegant infinite products.

Theorem 5.1. For $|q| < 1$,

$$(5.1)\quad \Phi_k(q) = \frac{\sum\limits_{m_1,\ldots,m_{k-1}=-\infty}^{\infty} \zeta^{(k-1)m_1+(k-2)m_2+\ldots m_{k-1}} q^{Q(m_1,m_2,\ldots,m_{k-1})}}{\prod\limits_{n=1}^{\infty}(1-q^n)^k},$$

where $\zeta = \exp(2\pi i/(k+1))$ and

$$(5.2)\quad Q(m_1,m_2,\ldots,m_{k-1}) = m_1^2 + m_2^2 + \ldots + m_{k-1}^2 + \sum_{1\leq i<j\leq k-1} m_i m_j .$$

Proof.

To facilitate our work we must introduce some standard notation [34; p. 89]:

$$(5.3)\quad (A;q)_n = (A)_n = (1-A)(1-Aq)\ldots(1-Aq^{n-1}),$$

$$(5.4)\quad (A;q)_\infty = (A)_\infty = \lim_{n\to\infty} (A;q)_n .$$

From the General Principle of Section 3 we see that $\Phi_k(q)$ is the constant term in

$$(5.5)\quad G_k(z) = \prod_{n=0}^{\infty} (1+zq^{(n+1)}+\ldots+z^k q^{k(n+1)})(1+z^{-1}q^n+\ldots+z^{-k}q^{kn}).$$

Now

$$(5.6)\quad G_k(z) = \prod_{n=0}^{\infty} \frac{(1-z^{k+1}q^{(k+1)(n+1)})(1-z^{-k-1}q^{(k+1)(n+1)})}{(1-zq^{n+1})(1-z^{-1}q^n)}$$

$$= \prod_{j=1}^{k} (\zeta^j zq)_\infty (\zeta^{-j}z^{-1})_\infty$$

$$= \frac{1}{(q)_\infty^k} \prod_{j=1}^{k} \sum_{m_j=-\infty}^{\infty} (-1)^{m_j} q^{\binom{m_j+1}{2}} z_j^{m_j} \zeta^{jm_j} .$$

By requiring $m_1 + m_2 + \ldots + m_k = 0$ (i.e. $m_k = -m_1 - m_2 - \ldots - m_{k-1}$) we can easily pick out the constant term.
Hence

(5.7) $$\Phi_k(q) = \frac{1}{(q)_\infty^k} \sum_{m_1,\ldots,m_{k-1}=-\infty}^{\infty} \zeta^{m_1+2m_2+\ldots+(k-1)m_{k-1}-k(m_1+\ldots+m_{k-1})} \times q^{\binom{m_1+1}{2}+\ldots+\binom{m_{k-1}+1}{2}+\binom{-m_1-\ldots-m_{k-1}+1}{2}}$$

$$= \frac{1}{(q)_\infty^k} \sum_{m_1,\ldots,m_{k-1}=-\infty}^{\infty} \zeta^{-(k-1)m_1-(k-2)m_2-\ldots-m_{k-1}} \times q^{m_1^2+m_2^2+\ldots m_{k-1}^2 + \sum_{1\leq i<j\leq k-1} m_i m_j}$$

$$= \frac{1}{(q)_\infty^k} \sum_{m_1,\ldots,m_{k-1}=-\infty}^{\infty} \zeta^{(k-1)m_1+(k-2)m_2+\ldots+m_{k-1}} \times q^{m_1^2+m_2^2+\ldots+m_{k-1}^2+ \sum_{1\leq i<j\leq k-1} m_i m_j}$$

(wherein each m_i was replaced by $-m_i$).

This is the desired formula for $\Phi_k(q)$. □

Corollary 5.1. For $|q| < 1$,

(5.8) $$\Phi_1(q) = \prod_{n=1}^{\infty} \frac{1}{1-q^n} ,$$

(5.9) $$\Phi_2(q) = \prod_{n=1}^{\infty} \frac{1}{(1-q^n)(1-q^{12n-10})(1-q^{12n-9})(1-q^{12n-3})(1-q^{12n-2})} .$$

$$(5.10)\quad \Phi_3(q) = \prod_{n=1}^{\infty} \frac{(1-q^{12n-6})}{(1-q^{6n-5})(1-q^{6n-4})^2(1-q^{6n-3})^3(1-q^{6n-2})^2(1-q^{6n-1})(1-q^{12n})}.$$

Proof. Equation (5.8) follows from Theorem 5.1 by immediate inspection. For equation (5.9) we see that with $\zeta = e^{2\pi i/3}$

$$(5.11)\quad \Phi_2(q) = \frac{\sum_{m_1=-\infty}^{\infty} \zeta^{m_1} q^{m_1^2}}{(q)_\infty^2}$$

$$= \frac{(q^2;q^2)_\infty(-\zeta q;q^2)_\infty(-\zeta^{-1}q;q^2)_\infty}{(q)_\infty^2}$$

(by Jacobi's triple product [4; p. 21])

$$= \frac{(q^2;q^2)\prod_{n=0}^{\infty}(1-q^{2n+1}+q^{4n+2})}{(q)_\infty^2}$$

$$= \frac{(q^2;q^2)_\infty(-q^3;q^6)_\infty}{(q)_\infty^2\,(-q;q^2)_\infty}$$

$$= \prod_{n=1}^{\infty}\frac{1}{(1-q^n)(1-q^{12n-10})(1-q^{12n-9})(1-q^{12n-3})(1-q^{12n-2})}.$$

Finally for $\Phi_3(q)$ with $\zeta = e^{2\pi i/4} = i$,

$$(5.12)\quad \Phi_3(q) = \frac{\sum_{m_1,m_2=-\infty}^{\infty} i^{2m_1+m_2} q^{m_1^2+m_2^2+m_1m_2}}{(q)_\infty^3}$$

$$= \frac{\sum_{m_1,m_2=-\infty}^{\infty} i^{2m_1+2m_2} q^{m_1^2+4m_2^2+2m_1m_2}}{(q)_\infty^3}$$

(the (m_1,m_2) and $(-m_1,-m_2)$ term cancel if m_2 odd)

$$= \frac{\sum_{m_2=-\infty}^{\infty} q^{3m_2^2} \sum_{m_1=-\infty}^{\infty} (-1)^{m_1+m_2} q^{(m_1+m_2)^2}}{(q)_\infty^3}$$

$$= \frac{\sum_{m_2=-\infty}^{\infty} q^{3m_2^2} \sum_{m_1=-\infty}^{\infty} (-1)^{m_1} q^{m_1^2}}{(q)_\infty^3}$$

(shifting m_1 to m_1-m_2 in inner sum)

$$= \frac{(q^6;q^6)_\infty(-q^3;q^6)_\infty^2(q^2;q^2)_\infty(q;q^2)_\infty^2}{(q)_\infty^3}$$

(by Jacobi's triple product identity [4; p. 21]

$$= \frac{(q^6;q^6)_\infty(q^6;q^{12})_\infty^2(q^2;q^2)_\infty(q;q^2)_\infty^2}{(q)_\infty^3(q^3;q^6)_\infty^2}$$

$$= \prod_{n=1}^{\infty} \frac{(1-q^{12n-6})}{(1-q^{6n-5})(1-q^{6n-4})^2(1-q^{6n-3})^3(1-q^{6n-2})^2(1-q^{6n-1})(1-q^{12n})},$$

this concludes Corollary 5.1. □

Theorem 5.2 For $|q| < 1$,

$$\text{(5.13)} \qquad C\Phi_k(q) = \frac{\sum_{m_1,m_2,\ldots,m_{k-1}=-\infty}^{\infty} q^{Q(m_1,m_2,\ldots,m_{k-1})}}{\prod_{n=1}^{\infty} (1-q^n)^k},$$

where $Q(m_1,\ldots,m_{k-1})$ is defined in (5.2).

Proof. From the General Principle of Section 3 we see that $C\Phi_k(q)$ is the constant term in

$$\text{(5.14)} \qquad CG_k(z) = \prod_{n=0}^{\infty} (1+zq^{n+1})^k(1+z^{-1}q^n)^k.$$

Now

$$(5.15)\qquad CG_k(z) = \prod_{j=1}^{k} (zq)_\infty (z^{-1})_\infty$$

$$= \frac{1}{(q)_\infty^k} \prod_{j=1}^{k} \sum_{m_j=-\infty}^{\infty} (-1)^{m_j} q^{\binom{m_j+1}{2}} z^{m_j} .$$

Thus everything now proceeds exactly as in Theorem 5.1 with $\zeta = 1$. □

Corollary 5.2. For $|q| < 1$,

$$(5.16)\qquad C\Phi_1(q) = \prod_{n=1}^{\infty} \frac{1}{1-q^n},$$

$$(5.17)\qquad C\Phi_2(q) = \prod_{n=1}^{\infty} \frac{(1-q^{4n-2})}{(1-q^{2n-1})^4(1-q^{4n})} .$$

$$(5.18)\qquad C\Phi_3(q) = \prod_{n=1}^{\infty} \frac{(1-q^{12n})(1-q^{12n-6})^3}{(1-q^{6n-5})^5(1-q^{6n-1})^5(1-q^{4n})^2(1-q^{6n-3})^7}$$

$$+ 4q \prod_{n=1}^{\infty} \frac{(1-q^{12n})(1-q^{4n})}{(1-q^{12n-6})(1-q^{4n-2})(1-q^n)^3} .$$

Proof. Equation (5.16) is the same as (5.18) and is only included for completeness. As for (5.17),

$$(5.19)\qquad C\Phi_2(q) = \frac{\sum_{m_1=-\infty}^{\infty} q^{m_1^2}}{(q)_\infty^2}$$

$$= \frac{(q^2;q^2)_\infty(-q;q^2)_\infty^2}{(q)_\infty^2}$$

(by Jacobi's triple product [4; p. 21])

$$= \frac{(q^2;q^2)_\infty (q^2;q^4)_\infty^2}{(q)_\infty^2 (q;q^2)_\infty^2}$$

$$= \prod_{n=1}^{\infty} \frac{(1-q^{4n-2})}{(1-q^{2n-1})^4(1-q^{4n})} .$$

Finally

$$(5.20) \qquad C\Phi_3(q) = \frac{\sum_{m_1,m_2=-\infty}^{\infty} q^{m_1^2+m_2^2+m_1m_2}}{(q)_\infty^3}$$

$$= \frac{\sum_{m_1,m_2=-\infty}^{\infty} q^{m_1^2+4m_2^2+2m_1m_2} + \sum_{m_1,m_2=-\infty}^{\infty} q^{m_1^2+(2m_2+1)^2+m_1(2m_2+1)}}{(q)_\infty^3}$$

$$= \frac{\sum_{m_2=-\infty}^{\infty} q^{3m_2^2} \sum_{m_1=-\infty}^{\infty} q^{(m_1+m_2)^2} + \sum_{m_2=-\infty}^{\infty} q^{2m_2^2+3m_2+1} \sum_{m_1=-\infty}^{\infty} q^{(m_1+m_2)^2+m_1+m_2}}{(q)_\infty^3}$$

$$= \frac{\sum_{m_2=-\infty}^{\infty} q^{3m_2^2} \sum_{m_1=-\infty}^{\infty} q^{m_1^2} + q \sum_{m_2=-\infty}^{\infty} q^{3m_2^2+3m_2} \sum_{m_1=-\infty}^{\infty} q^{m_1^2+m_1}}{(q)_\infty^3}$$

(wherein we have replaced m_1 by m_1-m_2)

$$= \frac{(q^6;q^6)_\infty(-q^3;q^6)_\infty^2(q^2;q^2)_\infty(-q;q^2)_\infty^2}{(q)_\infty^3}$$

$$+ \frac{4q(q^6;q^6)_\infty(-q^6;q^6)_\infty^2(q^2;q^2)_\infty(-q^2;q^2)_\infty^2}{(q)_\infty^3}$$

$$= \prod_{n=1}^{\infty} \frac{(1-q^{12n})(1-q^{12n-6})^3}{(1-q^{6n-5})^5(1-q^{6n-1})^5(1-q^{4n})^2(1-q^{6n-3})^7}$$

$$+ 4q \prod_{n=1}^{\infty} \frac{(1-q^{12n})(1-q^{4n})}{(1-q^{12n-6})(1-q^{4n-2})(1-q^n)^3} ,$$

and the proof of Corollary 5.2 is completed. □

The device used in proving (5.18) can obviously be iterated to get the expressions in (5.13) and (5.1) as sums of explicit infinite products. However, after the above results, the expressions quickly become long and messy.

6. Finite Generating Functions.

In the study of ordinary partitions it is often the case that polynomial generating functions play an important role. In particular, the Gaussian polynomial or q-binomial coefficients are studied extensively; these are defined by

$$(6.1) \qquad \begin{bmatrix} A \\ B \end{bmatrix}_q = \begin{cases} \dfrac{(1-q^A)(1-q^{A-1})\dots(1-q^{A-B+1})}{(1-q^B)(1-q^{B-1})\dots(1-q)} , & A \geqq 0,\ B \geqq 0 \\ 0, & \text{otherwise.} \end{cases}$$

Now these rational functions are in fact polynomials. Furthermore

$$(6.2) \qquad \begin{bmatrix} m+n \\ n \end{bmatrix}_q = \sum_{j=0}^{mn} p(m,n,j)q^j .$$

where $p(m,n,j)$ is the number of partitions of j into at most m parts each $\leqq n$.

One may also use the terminology of F-partitions. We can also say that $p(m,n,j)$ is the number of F-partitions of j with distinct parts in each row wherein each top entry is $< m$ and each bottom entry is $< n$. The finite form of Jacobi's Triple Product Identity [4; p. 49, Ex. 1] due to Cauchy:

$$(6.3) \qquad \prod_{j=1}^{m} (1+aq^j) \prod_{k=0}^{n-1} (1+z^{-1}q^k) = \sum_{j=-n}^{m} \begin{bmatrix} m+n \\ n+j \end{bmatrix}_q z^j q^{\binom{j+1}{2}}$$

may be used with the General Principle of Section 3 to establish that the constant term (i.e. $j = 0$) is in fact the generating function described by equation (6.2)

Since q-binomial coefficients have played such an important role in many branches of mathematics besides the theory of partitions, it does seem appropriate to introduce the corresponding objects for various classes of F-partitions.

Definition 6.1. Let $\phi_r(m,n,j)$ denote the number of those F-partitions enumerated by $\phi_r(j)$ (see (4.1)) with each entry in the top row $< m$ and each entry in the bottom row $< n$.

Definition 6.2. Let $c\phi_r(m,n,j)$ denote the number of those F-partitions enumerated by $c\phi_r(j)$ (see (4.8)) with each entry (regardless of subscript) in the top row $< m$ and each entry (regardless of subscript) in the bottom row $< n$.

We define the corresponding generating fucntions as follows

$$(6.4)\qquad {}_{rk}\begin{bmatrix} m+n \\ n \end{bmatrix}_q = \sum_{j\geq 0} \phi_k(m,n,j)q^j,$$

$$(6.5)\qquad {}_{ck}\begin{bmatrix} m+n \\ n \end{bmatrix}_q = \sum_{j\geq 0} c\phi_k(m,n,j)q^j.$$

The General Principle tells us immediately that ${}_{rk}\begin{bmatrix} m+n \\ n \end{bmatrix}_q$ is the constant term in

$$(6.6)\qquad \prod_{j=1}^{m} (1+zq^j+\ldots+z^kq^{kj}) \prod_{h=0}^{n-1} (1+z^{-1}q^h+\ldots+z^{-k}q^{hk}),$$

while ${}_{ck}\begin{bmatrix} m+n \\ n \end{bmatrix}_q$ is the constant term in

$$(6.7)\qquad \prod_{j=1}^{m} (1+zq^j)^k \prod_{h=0}^{n-1} (1+z^{-1}q^h)^k.$$

Obviously

$$(6.8)\qquad {}_{r1}\begin{bmatrix} m+n \\ n \end{bmatrix}_q = {}_{c1}\begin{bmatrix} m+n \\ n \end{bmatrix}_q = \begin{bmatrix} m+n \\ n \end{bmatrix}_q ,$$

the original Gaussian polynomial.

Also ${}_{rk}\begin{bmatrix} m+n \\ n \end{bmatrix}_q$ and ${}_{ck}\begin{bmatrix} m+n \\ n \end{bmatrix}_q$ are symmetric in m and n; this follows immediately by noting that the substitution $z \to z^{-1}q^{-1}$ does not change the constant term in either (6.6) or (6.7). The exact same procedures used in Section 5 (with (6.3) used instead of Jacobi's triple product identity) yield the following representations of these polynomials in terms of ordinary Gaussian polynomials. Since the proofs are essentially like those in Section 5, they are omitted.

Theorem 6.1.

$$(6.9)\qquad {}_{rk}\begin{bmatrix} m+n \\ n \end{bmatrix}_q = \sum_{h_1=-m}^{n} \cdots \sum_{h_{k-1}=-m}^{n} \begin{bmatrix} m+n \\ m+h_1 \end{bmatrix}_q \cdots \begin{bmatrix} m+n \\ m+h_{k-1} \end{bmatrix}_q$$

$$\times \begin{bmatrix} m+n \\ n+h_1+\dots+h_{k-1} \end{bmatrix}_q \zeta^{(k-1)h_1+(k-2)h_2+\dots+h_{k-1}} q^{Q(h_1,h_2,\dots,h_{k-1})}$$

$$(6.10)\qquad {}_{ck}\begin{bmatrix} m+n \\ n \end{bmatrix}_q = \sum_{h_1=-m}^{n} \cdots \sum_{h_{k-1}=-m}^{n} \begin{bmatrix} m+n \\ m+h_1 \end{bmatrix}_q \cdots \begin{bmatrix} m+n \\ m+h_{k-1} \end{bmatrix}_q$$

$$\times \begin{bmatrix} m+n \\ n+h_1+\dots+h_{k-1} \end{bmatrix}_q q^{Q(h_1,h_2,\dots,h_{k-1})} ,$$

where $Q(h_1,h_2,\dots,h_{k-1})$ is the quadratic form defined in (5.2).

It appears that there are not significant simplifications of the formulas in (6.9) and (6.10). For example, the F-partitions enumerated by ${}_{r2}\begin{bmatrix} 4 \\ 2 \end{bmatrix}_q$ are

$$\phi,\ \begin{pmatrix} 0 \\ 0 \end{pmatrix},\ \begin{pmatrix} 00 \\ 00 \end{pmatrix},\ \begin{pmatrix} 10 \\ 00 \end{pmatrix},\ \begin{pmatrix} 00 \\ 10 \end{pmatrix},\ \begin{pmatrix} 1 \\ 0 \end{pmatrix},\ \begin{pmatrix} 0 \\ 1 \end{pmatrix},$$

$$\binom{1}{1},\ \binom{11}{00},\ \binom{00}{11},\ \binom{10}{10},\ \binom{100}{100},\ \binom{110}{100},$$
$$\binom{100}{110},\ \binom{11}{10},\ \binom{10}{11},\ \binom{11}{11},\ \binom{110}{110},\ \binom{1100}{1100},$$

Hence

$$_{r2}\begin{bmatrix} 4 \\ 2 \end{bmatrix}_q = 1 + q + 3q^2 + 3q^3 + 3q^4 + 3q^5 + 3q^6 + q^7 + q^8 .$$

From (6.9), we see that

$$_{r2}\begin{bmatrix} 4 \\ 2 \end{bmatrix}_q = (\zeta^2+\zeta^{-2})q^4 + (\zeta+\zeta^{-1})q \begin{bmatrix} 4 \\ 1 \end{bmatrix}_q^2 + \begin{bmatrix} 4 \\ 2 \end{bmatrix}_q^2$$

$$= -q^4 - q \begin{bmatrix} 4 \\ 1 \end{bmatrix}_q^2 + \begin{bmatrix} 4 \\ 2 \end{bmatrix}_q^2$$

$$= -q^4 - q(1+q+q^2+q^3)^2 + (1+q+2q^2+q^3+q^4)^2$$

$$= 1 + q + 3q^2 + 3q^3 + 3q^4 + 3q^5 + 3q^6 + q^7 + q^8 .$$

The polynomials $_{r2}\begin{bmatrix} m+n \\ n \end{bmatrix}_q$ and $_{c2}\begin{bmatrix} m+n \\ n \end{bmatrix}_q$ could conceivably be important in the considerations of the conjectures in Section 11. For this reason we explicitly exhibit their particular cases as the following result:

Corollary 6.1.

$$(6.11)\qquad _{r2}\begin{bmatrix} m+n \\ n \end{bmatrix}_q = \sum_{h=-m}^{n} \begin{bmatrix} m+n \\ m+h \end{bmatrix}_q \begin{bmatrix} m+n \\ n+h \end{bmatrix}_q \zeta^h q^{h^2} ;$$

$$(6.12)\qquad _{c2}\begin{bmatrix} m+n \\ n \end{bmatrix}_q = \sum_{h=-m}^{n} \begin{bmatrix} m+n \\ m+h \end{bmatrix}_q \begin{bmatrix} m+n \\ n+h \end{bmatrix}_q q^{h^2} ,$$

where $\zeta = \exp(2\pi i/3)$.

Finally we remark that (6.9) is slightly less satisfactory than (6.10) in that it appears to have complex numbers (rather than reals) as coefficients. The realness of the coefficients is clear in this next result.

Theorem 6.2.

$$(6.13)\qquad {}_{rk}\begin{bmatrix} m+n \\ n \end{bmatrix}_q = \sum_{r\geq 0} \begin{bmatrix} m+n \\ r \end{bmatrix}_{q^{k+1}} \begin{bmatrix} m+(n-r)(k+1)-1 \\ m+n-1 \end{bmatrix}_q q^{(k+1)\binom{r}{2}-k\binom{n}{2}} (-1)^r$$

Proof. The polynomial ${}_{rk}\begin{bmatrix} m+n \\ n \end{bmatrix}_q$ is the constant term in

$$(6.14)\qquad \prod_{j=1}^{m} (1+zq^j+z^2q^{2j}+\ldots+z^kq^{kj}) \prod_{h=0}^{n-1} (1+z^{-1}q^h+\ldots+z^{-k}q^{kh})$$

$$= \prod_{j=1}^{m} \frac{(1-z^{k+1}q^{(k+1)j})}{(1-zq^j)} \prod_{h=0}^{n-1} \frac{(1-z^{-k-1}q^{(k+1)h})}{(1-z^{-1}q^h)}$$

$$= z^{-kn}q^{k\binom{n}{2}} \prod_{t=0}^{m+n-1} \frac{(1-z^{k+1}q^{-(k+1)(n-1)+(k+1)t})}{(1-zq^{-n+1+t})}$$

$$= z^{-kn}q^{k\binom{n}{2}} \sum_{r=0}^{m+n} \begin{bmatrix} m+n \\ r \end{bmatrix}_{q^{k+1}} (-1)^r z^{r(k+1)} q^{(k+1)\binom{r+1}{2}-nr(k+1)}$$

$$\times \sum_{s=0}^{\infty} \begin{bmatrix} m+n+s-1 \\ s \end{bmatrix}_q z^s q^{s(1-n)}$$

(by [4; p. 36, Th. 313]).

Hence the constant term arises when $s = kn - r(k+1)$, and this leads immediately to (6.13). □

The disadvantage of Theorem 6.2 is that the series is not obviously a polynomial q. The case $k = 1$ provides a series that sums to the ordinary Gaussian polynomial; it is in fact (when $k = 1$) a special case of the q-Pfaff-Saalschutz summation [4; p. 37, eq. (3.3.11)].

7. Hecke Modular Forms.

So far we have obtained several representations for the generating functions related to F-partitions. These have related to multi-dimensional theta functions (Section 5) and their finite analogs (Section 6). Surprisingly the $\Phi_k(q)$ (however, unfortunately, not the $C\Phi_k(q)$) have very nice representations in terms of Hecke modular forms. In a recent paper [7], the following lemma was proved concerning Hecke modular forms:

Lemma [7]. The coefficient of z^0 in the Laurent series expansion of

$$(7.1)\qquad \prod_{n=1}^{\infty} \frac{(1-q^n)^2(1-q^{Bn})(1-z^A q^{B(n-1)+C})(1-z^{-A}q^{Bn-C})}{(1-zq^n)(1-z^{-1}q^{n-1})}$$

(wherein $|q| < 1$; A,B,C are positive integers) in the annulus $1 < |z| < |q|^{-1}$ is

$$(7.2)\qquad \sum_{\substack{j,r=-\infty \\ r\geq A|j|}}^{\infty} (-1)^{r+(A+1)j} q^{((B-A^2)j^2+r^2)/2 + ((B+A-2C)j+r)/2} .$$

From this lemma (proved in Section 2 of [7]) we may deduce

Theorem 7.1.

$$(7.3)\qquad \Phi_k(q) = \frac{\displaystyle\sum_{\substack{j,r=-\infty \\ r\geq(k+1)|j|}}^{\infty} (-1)^{r+kj} q^{\binom{r+1}{2}-\binom{k+1}{2}j^2}}{\displaystyle\prod_{n=1}^{\infty}(1-q^n)^2(1-q^{(k+1)n})} .$$

Proof. From (5.4) we recognize $\{\prod_{n\geq1}(1-q^n)^2(1-q^{(k+1)n})\}\Phi_k(q)$ as the constant term in (7.1) when $A = B = C = k + 1$. Our result now follows from the above lemma. □

Now the series in the numerator of (7.3) is both real and highly convergent. This allows one to create quite efficient algorithms for computing $\phi_k(n)$. One merely needs to compute the coefficients in $\prod_{n=1}^{\infty}(1-q^n)^{-2}(1-q^{(k+1)n})^{-1}$ using the devices of [4; p. 98, Ex. 2]. The numerator on the right of (7.3) is fairly sparse since truncating the j-series at $|j| = N$ introduces an error $O(q^{(k+1)\binom{N+1}{2}})$.

8. q-Series.

So far we have looked primarily at representations of various generating functions in terms of theta series, infinite products and finite approximations thereof. It is the case, however, that many useful properties of partitions arise from q-series or Eulerian series expansions. Indeed the following famous result of Jacobi [25; p. 233] (often attributed to Euler) has proved very important in several contexts [24; pp. 87-88], [21; pp. 871-872], [22; pp. 91-92]:

$$(8.1)\qquad 1+\prod_{n=1}^{\infty}\frac{q^{n^2}}{(1-q)^2(1-q^2)^2\cdots(1-q^n)^2}=\prod_{n=1}^{\infty}\frac{1}{(1-q^n)}\,.$$

Actually the r-th term in the series (8.1) is generating function for the number of F-partitions enumerated by $\phi_1(n)$ with r entries in each row [4; pp. 27-28].

The most natural extensions of (8.1) to all classes of F-partitions under consideration seem to be embodied in the following two theorems:

Theorem 8.1.

$$(8.2)\qquad \Phi_k(q)=\sum_{r=0}^{\infty}\sum_{s=0}^{\infty}\sum_{t=0}^{\infty}\frac{(-1)^{r+s}q^{(k+1)\binom{r+1}{2}+(k+1)\binom{s}{2}+t}}{(q)_t(q^{k+1};q^{k+1})_r(q^{k+1};q^{k+1})_s(q)_{(r-s)(k+1)+t}}$$

Proof. We recall from the proof of Theorem 5.1 that $\Phi_k(q)$ is the constant term in

$$\prod_{j=1}^{\infty}(1+zq^j+z^2q^{2j}+\ldots+z^kq^{kj})\prod_{h=0}^{\infty}(1+z^{-1}q^h+\ldots+z^{-k}q^{kh})$$

$$=\prod_{j=1}^{\infty}\frac{(1-z^{k+1}q^{(k+1)j})}{(1-zq^j)}\prod_{h=0}^{\infty}\frac{(1-z^{-k-1}q^{(k+1)h})}{(1-z^{-1}q^h)}$$

$$=\sum_{r=0}^{\infty}\frac{(-1)^rz^{r(k+1)}q^{(k+1)\binom{r+1}{2}}}{(q^{k+1};q^{k+1})_r}\sum_{t=0}^{\infty}\frac{z^tq^t}{(q)_t}$$

$$\times\sum_{s=0}^{\infty}\frac{(-1)^sz^{-s(k+1)}q^{(k+1)\binom{s}{2}}}{(q^{k+1};q^{k+1})_s}\sum_{u=0}^{\infty}\frac{z^{-u}}{(q)_u}\,.$$

The constant term is derived by setting $u=(r-s)(k+1)+t$, and the desired result follows. □

It should be noted that standard transformations from the theory of basic hypergeometric series can be utilized to transform (8.2) into (7.3). In fact, this discovery eventually led to the general result of [7].

Theorem 8.2. For $|q|<1$,

$$(8.3)\quad C\Phi_k(q) = \frac{1}{(q)_\infty} \sum_{n_1=0}^{\infty} \cdots \sum_{n_{k-1}=0}^{\infty} \sum_{m_1=0}^{\infty} \cdots \sum_{m_{k-1}=0}^{\infty} \frac{q^{R(n_1,\ldots,n_{k-1},m_1,\ldots,m_{k-1})}}{(q)_{n_1}\cdots(q)_{n_{k-1}}(q)_{m_1}\cdots(q)_{m_{k-1}}}$$

where R is the quadratic form given by

$$(8.4)\quad R(n_1,\ldots,n_{k-1},m_1,\ldots,m_{k-1}) = \sum_{i=1}^{k-1} (n_i^2+m_i^2) + \sum_{1\le i<j\le k-1} (n_in_j+m_im_j) - \sum_{1\le i,j\le k-1} n_im_j .$$

Proof. From (5.14) we see that $C\Phi_k(q)$ is the constant term in

$$(8.5)\quad \prod_{n=1}^{\infty} (1+zq^n)^k(1+z^{-1}q^{n-1})^k$$

$$= \sum_{n_1,\ldots,n_k\ge 0} \sum_{m_1,\ldots,m_k\ge 0} \frac{z^{n_1+\cdots+n_k-m_1-\cdots-m_k} q^{\binom{n_1+1}{2}+\cdots+\binom{n_k+1}{2}+\binom{m_1}{2}+\cdots+\binom{m_k}{2}}}{(q)_{n_1}\cdots(q)_{n_k}(q)_{m_1}\cdots(q)_{m_k}}$$

Hence the constant term arises for $m_k = n_1 + \cdots + n_k - m_1 - \cdots - m_{k-1}$, and so

$$(8.6)\quad C\Phi_k(q) = \sum_{n_1,\ldots,n_k\ge 0} \sum_{m_1,\ldots,m_{k-1}\ge 0} \frac{q^{\binom{n_1+1}{2}+\cdots+\binom{n_k+1}{2}+\binom{m_1}{2}+\cdots+\binom{m_{k-1}}{2}}}{(q)_{n_1}\cdots(q)_{n_k}(q)_{m_1}\cdots(q)_{m_{k-1}}}$$

$$\times \frac{q^{\binom{n_1+\cdots+n_k}{2}+\binom{m_1+\cdots+m_{k-1}+1}{2}-(n_1+\cdots+n_k)(m_1+\cdots+m_{k-1})}}{(q)_{n_1+\cdots+n_k-m_1-\cdots-m_{k-1}}} .$$

Now we know that for any integer A positive, negative or zero [4; p. 20, eq. (2.2.8)]

$$(8.7)\quad \sum_{n=0}^{\infty} \frac{q^{n(n+A)}}{(q)_n(q)_{n+A}} = \frac{1}{(q)_\infty} ,$$

where we note $(q)_n^{-1} = 0$ if $n < 0$. Thus we may use (8.7) on the n_k sum with $A = n_1+\cdots+n_{k-1}-m_1-\cdots-m_{k-1}$. Hence

$$(8.8)\qquad C\Phi_k(q) = \frac{1}{(q)_\infty} \sum_{n_1,\dots,n_{k-1}\geq 0} \sum_{m_1,\dots,m_{k-1}\geq 0} \frac{q^{n_1^2+\dots+n_{k-1}^2+m_1^2+\dots+m_{k-1}^2+\sum_{1\leq i<j\leq k-1}(n_in_j+m_im_j)-\sum_{1\leq i,j\leq k-1} n_im_j}}{(q)_{n_1}\cdots(q)_{n_{k-1}}(q)_{m_1}\cdots(q)_{m_{k-1}}}$$

as desired. □

Corollary 9.1. For $|q| < 1$

$$(8.9)\qquad \sum_{n_1,\dots,n_{k-1}\geq 0} \sum_{m_1,\dots,m_{k-1}\geq 0} \frac{q^{R(n_1,\dots,n_{k-1},m_1,\dots,m_{k-1})}}{(q)_{n_1}\cdots(q)_{n_{k-1}}(q)_{m_1}\cdots(q)_{m_{k-1}}}$$

$$= \frac{\sum_{s_1,\dots,s_{k-1}=-\infty}^{\infty} q^{Q(s_1,\dots,s_{k-1})}}{(q)_\infty^{k-1}},$$

where R is defined by (8.4) and Q is defined by (5.2).

Proof. We identify the right sides of (5.13) and (8.3) and multiply through by $(q)_\infty$.

The case $k = 2$ of the above result forms a crucial identity in the treatment [5], [19] of the Lusztig-Macdonald-Wall Conjectures. For this reason we record it separately.

Corollary 8.2 [5; p. 92]. For $|q| < 1$,

$$(8.10)\qquad \sum_{n,m\geq 0} \frac{q^{n^2+m^2-nm}}{(q)_n(q)_m} = \frac{\sum_{s=-\infty}^{\infty} q^{s^2}}{(q)_\infty}.$$

Proof. This is the case $k = 2$ of Corollary 8.1. □

Unfortunately the roots of unity required in (5.6) do not allow the utilization of (8.7) for a treatment of $\Phi_k(q)$ similar to Theorem 8.2.

It is also possible to prove an analog of the q-Vandermonde convolution

[4; p. 37, eq. (3.3.10)] for $\begin{bmatrix} m+n \\ n \end{bmatrix}_{ck\ q}$. One merely follows the proof of Theorem 8.2 but starts with

$$(8.11)\qquad \prod_{j=1}^{m} (1+zq^j)^k \prod_{h=0}^{n-1} (1+z^{-1}q^h)^k ,$$

instead of (8.5). The result that follows is

$$(8.12)\qquad {}_{ck}\begin{bmatrix} m+n \\ n \end{bmatrix}_q = \sum_{n_1=0}^{m} \cdots \sum_{n_{k-1}=0}^{m} \sum_{m_1=0}^{n} \cdots \sum_{m_1=0}^{n} q^{R(n_1,\ldots,n_{k-1},m_1,\ldots,m_{k-1})}$$

$$\times \begin{bmatrix} m \\ n_1 \end{bmatrix}_q \cdots \begin{bmatrix} m \\ n_{k-1} \end{bmatrix}_q \begin{bmatrix} n \\ m_1 \end{bmatrix}_q \cdots \begin{bmatrix} n \\ m_{k-1} \end{bmatrix}_q \begin{bmatrix} m+n \\ m+n_1+\ldots+n_{k-1}-m_1-\ldots-m_{k-1} \end{bmatrix}_q$$

We conclude this section by noting the possibility of utilizing the classical Rogers-Ramanujan type identities collected by L. J. Slater [33] to find further q-series representation of our generating functions. For example:

Theorem 8.3. For $|q| < 1$,

$$(8.13)\qquad \Phi_2(-q) = \frac{(q;q^2)_\infty}{(-q;q^2)_\infty^3 (q^4;q^4)_\infty} \left\{ 1+3 \sum_{n=1}^{\infty} \frac{(q^3;q^3)_{n-1}(-q)_n q^{\binom{n+1}{2}}}{(q)_{2n}(q)_{n-1}} \right\}$$

Proof. By (5.9), we know that

$$(8.14)\qquad \Phi_2(q) = \frac{1}{(q)_\infty^2} \prod_{n=1}^{\infty} (1-q^{6n})(1-q^{6n-1})(1-q^{6n-5})(1-q^{12n-8})(1-q^{12n-4})$$

$$= \frac{1}{(q)_\infty^2} \left(\sum_{n=-\infty}^{\infty} q^{9n^2} - q \sum_{n=-\infty}^{\infty} q^{9n^2+6n} \right)$$

(by [3; p. 466, Th. 3.9])

$$= \frac{(q^{18};q^{18})_\infty (-q^9;q^{18})_\infty^2}{(q)_\infty^2} +$$

$$- \frac{q(-q^3;q^{18})_\infty(-q^{15};q^{18})_\infty(q^{18};q^{18})_\infty}{(q)_\infty^2}$$

Now by [33; p. 160, eq. (78)]

$$(8.15)\qquad (q^{18};q^{18})_\infty(q^9;q^{18})_\infty^2$$

$$= \frac{(q)_\infty}{(-q)_\infty}\left\{1+2\sum_{n=1}^{\infty}\frac{(q^3;q^3)_{n-1}(-q)_n q^{\binom{n+1}{2}}}{(q)_{2n}(q)_{n-1}}\right\},$$

and by [33; p. 160, eq. (76)]

$$(8.16)\qquad q(q^3;q^{18})_\infty(q^{15};q^{18})_\infty(q^{18};q^{18})_\infty$$

$$= \frac{(q)_\infty}{(-q)_\infty}\sum_{n=1}^{\infty}\frac{(q^3;q^3)_{n-1}(-q)_n q^{\binom{n+1}{2}}}{(q)_{2n}(q)_{n-1}}.$$

If we change the sign of q in (8.14) and then substitute in the expressions in (8.15) and (8.16), we find

$$(8.17)\qquad \Phi_2(-q) = \frac{1}{(-q;-q)_\infty^2}\,\frac{(q)_\infty}{(-q)_\infty}\left\{1+3\sum_{n=1}^{\infty}\frac{(q^3;q^3)_{n-1}(-q)_n q^{\binom{n+1}{2}}}{(q)_{2n}(q)_{n-1}}\right\}.$$

Equation (8.13) follows once we observe that

$$\frac{1}{(-q;-q)_\infty^2}\,\frac{(q)_\infty}{(-q)_\infty} = \frac{(q;q^2)_\infty(q^2;q^2)_\infty}{(-q;q^2)_\infty^2(q^2;q^2)_\infty^2(-q;q^2)_\infty(-q^2;q^2)_\infty}$$

$$= \frac{(q;q^2)}{(-q;q^2)_\infty^3(q^4;q^4)_\infty}. \qquad \square$$

9. Kloosterman Quadratic Forms.

H. D. Kloosterman [26] was the first to study systematically the number $r_k(n,m)$ defined to be the number of simultaneous solutions in integers of

$$(9.1)\qquad x_1^2 + x_2^2 + \dots + x_k^2 = n, \qquad x_1 + x_2 + \dots + x_k = m.$$

Subsequent authors [11], [28], [39] both simplified and extended his results. Kloosterman [26] found divisor sum formulas for $r_k(n,m)$ with $k = 3,5$ and 7 that are as striking as the classical divisor sum formulas of Jacobi for 4, 6 and 8 squares [25],[35; pp. 306-307].

We note that n must be congruent to $m \pmod 2$ for (9.1) to have any solutions at all. Furthermore $r_k(2n,0)$ is the number of solutions of

$$n = \{(x_1+x_2+\ldots+x_{k-1})^2 + x_1^2 + x_2^2 + \ldots + x_{k-1}^2\}/2 = Q(x_1,x_2,\ldots,x_{k-1}), \tag{9.2}$$

the quadratic form defined in (5.2). This observation applied to Theorem 5.2 shows immediately that:

Theorem 9.1. For $|q| < 1$,

$$C\Phi_k(q) = \frac{\sum_{n=0}^{\infty} r_k(2n,0)q^n}{\prod_{n=1}^{\infty}(1-q^n)^k}. \tag{9.3}$$

Now of course the cases $k = 1$ and 2 cannot have nicer representations than those given in Theorem 5.2 and Corollary 5.2. However we may utilize Kloosterman's results for $k = 3,5$ to provide further elegant formulas for the $C\Phi_k(q)$. We omit $k = 7$ since it is similar to $k = 5$.

Corollary 9.1. For $|q| < 1$,

$$C\Phi_3(q) = \frac{1+6\sum_{n=0}^{\infty}\left(\frac{q^{3n+1}}{1-q^{3n-1}} - \frac{q^{3n+2}}{1-q^{3n+2}}\right)}{\prod_{n=1}^{\infty}(1-q^n)^3}; \tag{9.4}$$

$$C\Phi_5(q) = \frac{1 + \sum_{n=1}^{\infty} q^n 25n\left(1 - \left(\frac{n_0}{5}\right)5^{-\beta}\right)\sum_{d|n}\left(\frac{d}{5}\right)\frac{1}{d}}{\prod_{n=1}^{\infty}(1-q^n)^5}, \tag{9.5}$$

(where $n = 5^{\beta-1}n_0$, $(n_0,5) = 1$, $\left(\frac{d}{5}\right)$ is the Legendre symbol),

Proof. By Kloosterman's Hauptsatz III [26; p. 362] we see that for $n > 0$

$$r_3(2n,0) = 6\sum_{d|n}\left(\frac{d}{3}\right), \tag{9.6}$$

where $(\frac{d}{3})$ is the Legendre symbol defined as 0 if $3|d$ and ±1 if $d \equiv \pm1 \pmod 3$. Hence

$$(9.7)\qquad C\Phi_3(q) = \frac{1}{(q)_\infty^3}\left(1+6\sum_{n=1}^{\infty} q^n \sum_{d|n} (\tfrac{d}{3})\right)$$

$$= \frac{1}{(q)_\infty^3}\left(1+6\sum_{r=1}^{\infty}\sum_{s=1}^{\infty} q^{rs} (\tfrac{r}{3})\right)$$

$$= \frac{1}{(q)_\infty^3}\left(1+6\sum_{r=1}^{\infty}\frac{(\frac{r}{3})q^r}{1-q^r}\right)$$

$$= \frac{1}{(q)_\infty^3}\left(1+6\sum_{n=0}^{\infty}\left(\frac{q^{3n+1}}{1-q^{3n+1}} - \frac{q^{3n+2}}{1-q^{3n+2}}\right)\right).$$

Identity (9.5) follows immediately from Theorem 9.1 and Kloosterman's Hauptsatz I [26; p. 358]. This last result could, of course, be translated into a purely q-series identity just as (9.6) led to (9.4). However the result appears sufficiently complicated that we have chosen to leave the divisor sum in (9.5).

10. Congruences.

There are a number of interesting congruences for both the $\phi_k(n)$ and $c\phi_k(n)$. Some resemble the familiar property of $p(n)$ $(= \phi_1(n) = c\phi_1(n))$ due to Ramanujan [31; p. 210]:

$$(10.1)\qquad p(5n+4) \equiv 0 \pmod 5.$$

Also there is an infinite family of results for $c\phi_k(n)$ which is most easily established in a purely combinatorial manner. We begin with the analogs of (10.1).

Theorem 10.1. Let $\{a_n\}_{n=0}^{\infty}$ be a sequence of integers. Then all coefficients of q^{5n+3} for $n = 0,1,2,...$ in

$$(10.2)\qquad \frac{\sum_{m=0}^{\infty} a_m q^{m^2}}{\prod_{n=1}^{\infty}(1-q^n)^2},$$

are divisible by 5.

Proof. Let $f(q)$ denote the expression in (10.2). Then

$$f(q) = \frac{\prod_{n=1}^{\infty}(1-q^n)^3 \sum_{m=0}^{\infty} a_m q^{m^2}}{\prod_{n=1}^{\infty}(1-q^n)^5}$$

$$\equiv \frac{\sum_{j=0}^{\infty}(-1)^j(2j+1)q^{\binom{j+1}{2}} \sum_{m=0}^{\infty} a_m q^{m^2}}{\prod_{n=1}^{\infty}(1-q^{5n})} \pmod 5.$$

(by [4; p. 176, Ex. 7])

The theorem is then proved provided we now show that all coefficients of q^{5n+3} in the numerator of this last expression are divisible by 5.

In order to get a contribution to q^{5n+3}, we must have

$$\frac{j(j+1)}{2} + m^2 \equiv 3 \pmod 5$$

or equivalently

$$(2j+1)^2 + 8m^2 \equiv 0 \pmod 5,$$

but the only solutions of this are

$$2j + 1 \equiv m \equiv 0 \pmod 5.$$

Hence each contribution $(-1)^j(2j+1)a_m$ to the coefficient of q^{5n+3} is divisible by 5. Thus Theorem 10.1 is established. □

Corollary 10.1.

(10.3) $\phi_2(5n+3) \equiv c\phi_2(5n+3) \equiv 0 \pmod 5.$

Proof. First we note from (5.9) that

$$(10.4)\qquad \Phi_2(q) = \prod_{n=1}^{\infty} \frac{(1-q^{6n-5})(1-q^{6n-1})(1-q^{6n})(1-q^{12n-4})(1-q^{12n-8})}{(1-q^n)^2}$$

$$= \frac{\sum_{m=-\infty}^{\infty} q^{9m^2} - \sum_{m=-\infty}^{\infty} q^{9m^2+6m+1}}{\prod_{n=1}^{\infty}(1-q^n)^2}$$

(by the quintuple product identity [3; p. 466, Th. 3.9])

$$= \frac{\sum_{m=0}^{\infty} a_m q^{m^2}}{\prod_{n=1}^{\infty}(1-q^n)^2},$$

where $a_0 = 1$, $a_m = 2$ if $3|m$, $a_m = -1$ if $3\nmid m$. Thus by Theorem 10.1, we see that $\phi_2(5n+3) \equiv 0 \pmod 5$.

By (5.13) with $k = 2$, we see that

$$(10.5)\qquad C\Phi_2(q) = \frac{\sum_{m=-\infty}^{\infty} q^{m^2}}{\prod_{n=1}^{\infty}(1-q^n)^2} = \frac{1+2\sum_{m=1}^{\infty} q^{m^2}}{\prod_{n=1}^{\infty}(1-q^n)^2},$$

and again Theorem 10.1 implies that $c\phi_2(5n+3) \equiv 0 \pmod 5$. □

We now move on to congruences that are only applicable to $c\phi_k(n)$.

Theorem 10.2. If k is a prime and $k\nmid n$, then

$$(10.6)\qquad c\phi_k(n) \equiv 0 \pmod{k^2}.$$

Proof. Let us call two F-partitions enumerated by $c\phi_k(n)$ equivalent if each may be obtained from the other by merely a change of colors on the parts. I.e. if $k = 3$ then

$$\begin{pmatrix} 0_3 & 0_1 \\ 0_3 & 0_1 \end{pmatrix}, \begin{pmatrix} 0_3 & 0_1 \\ 0_2 & 0_1 \end{pmatrix}, \begin{pmatrix} 0_3 & 0_1 \\ 0_3 & 0_2 \end{pmatrix}, \begin{pmatrix} 0_3 & 0_2 \\ 0_3 & 0_1 \end{pmatrix}, \begin{pmatrix} 0_3 & 0_2 \\ 0_2 & 0_1 \end{pmatrix},$$

(10.7)
$$\begin{pmatrix} 0_3 & 0_2 \\ 0_3 & 0_2 \end{pmatrix}, \begin{pmatrix} 0_2 & 0_1 \\ 0_3 & 0_1 \end{pmatrix}, \begin{pmatrix} 0_2 & 0_1 \\ 0_2 & 0_1 \end{pmatrix}, \begin{pmatrix} 0_2 & 0_1 \\ 0_3 & 0_2 \end{pmatrix}$$

are all equivalent.

Now we consider two equilength rows of uncolored nonnegative integers

$$\lambda_1\lambda_1\cdots\lambda_1\lambda_2\lambda_2\cdots\lambda_2\cdots\lambda_j\lambda_j\cdots\lambda_j$$

$$\mu_1\mu_1\cdots\mu_1\mu_2\mu_2\cdots\mu_2\cdots\mu_h\mu_h\cdots\mu_h$$

where λ_i appears $f_i > 0$ times, μ_i appears $g_i > 0$ and $f_1 + f_2 + \cdots + f_j = g_1 + g_2 + \cdots + g_h$. This array can be thought of as representing one equivalence class of F-partitions; all we have to do is color the parts (remember descending order of colors). This can be done in

(10.8) $$\left(\binom{k}{f_1}\binom{k}{f_2}\cdots\binom{k}{f_j}\right)\left(\binom{k}{g_1}\binom{k}{g_2}\cdots\binom{k}{g_h}\right)$$

ways.

Since k is prime and the f's and g's are positive the only way that the expression in (10.8) is not divisible by k^2 is if at most one of the f's and g's $\neq k$.

If all the f's and g's equal k, then the number n being partitioned equals

$$\sum_{i=1}^{j} f_i + \Sigma f_i\lambda_i + \Sigma g_i\mu_i$$

$$= kj + \Sigma k\lambda_i + \Sigma k\mu_i$$

$$= k(j + \Sigma\lambda_i + \Sigma\mu_i),$$

and $k|n$.

If exactly one of the f's and g's is not equal k, then it is an integral multiple of k because $f_1 + f_2 + \cdots + f_j = g_1 + g_2 + \cdots + g_h$. Thus if $g_\alpha \neq k$ is the particular one, then $g_\alpha > k$ and so $\binom{k}{g_\alpha} = 0$ and the expression in (10.8) is divisible by k^2 after all.

Thus as long as we are partitioning integers n that are relatively prime to k we see that all such partitions of n can be divided into equivalence classes for which the cardinality of each equivalence class is a multiple of k^2. Thus Theorem 10.2 is established. □

Corollary 10.2. For k prime there exists $G_k(q)$ analytic inside $|q| < 1$ with integral power series coefficients such that

$$(10.9) \qquad C\Phi_k(q) = \prod_{n=1}^{\infty} \frac{1}{1-q^{kn}} + k^2 G_k(q).$$

Proof. All the equivalence classes considered in the proof of Theorem 10.2 have multiples of k^2 elements except those singleton classes that have

$$f_1 = f_2 = \ldots = f_j = g_1 = g_2 = \ldots = g_h = k.$$

Thus these classes are produced by taking one of the original Frobenius symbols (i.e. an F-partition enumerated by $c\phi_1(n) = \phi_1(n) = p(n)$) and replicating each number on each row k times. Hence all of the singleton equivalence classes have as generating function

$$(10.10) \qquad C\Phi_1(q^k) = \sum_{n \geq 0} c\phi_1(n) q^{kn} = \prod_{n=1}^{\infty} \frac{1}{1-q^{kn}},$$

and Corollary 10.2 follows. □

Corollary 10.3. For k prime there exists polynomials $G_k(m,n;q)$ with integral coefficients such that

$$(10.11) \qquad {}_{ck}\begin{bmatrix} m+n \\ n \end{bmatrix}_q = \begin{bmatrix} m+n \\ n \end{bmatrix}_{q^k} + k^2 G_k(m,n;q).$$

Proof. Everything is the same as the proof of Corollary 10.2 except that top row entries are all $< m$ and bottom row entries are all $< n$. □

Theorem 10.2 can be inelegantly extended to composite k; by considering only k prime we can simply reveal how this type of argument works.

11. The Dyson Rank for F-partitions.

In 1944, F. J. Dyson [16] examined partitions relative to what he called the "rank" of a partition.

Definition 11.1. The D-rank of a partition π is the largest part minus the number of parts. If π is written as a Frobenius partition, i.e.

$$\begin{pmatrix} a_1 a_2 & \cdots & a_r \\ b_1 b_2 & \cdots & b_r \end{pmatrix} \qquad \begin{array}{l} (a_1 > a_2 \cdots > a_r \geqq 0, \\ \ b_1 > b_2 \cdots > b_r \geqq 0) \end{array}$$

then the D-rank of π is $a_1 - b_1$.

Definition 11.2. The D-rank of any generalized Frobenius partition

$$\begin{pmatrix} a_1 a_2 \ldots a_r \\ b_1 b_2 \ldots b_r \end{pmatrix}$$

is defined to be $a_1 - b_1$ (if the parts are colored, the D-rank is the numerical magnitude of $a_1 - b_1$).

We use the term D-rank because group theorists use rank to refer to the length of the rows [27; p. 60].

Dyson conjectured [16] and Atkin and Swinnerton-Dyler [9] proved the following combinatorial interpretation of Ramanujan's congruence (10.1):

Theorem. [9; p. 101]. Let $N(m,t,n)$ denote the number of partitions of n with D-rank congruent to m modulo t. Then

$$(11.1) \quad N(0,5,5n+4) = N(1,5,5n+4) = N(2,5,5n+4) = N(3,5,5n+4) = N(4,5,5n+4) = \frac{1}{5} p(5n+4).$$

While this is a purely combinatorial assertion that implies Ramanujan's congruence, it nonetheless requires a very intricate proof [9; Sections 5 and 6]. No purely combinatorial proof is known.

Definition 11.3. Let $N_k(m,t,n)$ (resp. $cN_k(m,t,n)$) denote the number of F-partitions enumerated by $\phi_k(n)$ (resp. $c\phi_k(n)$) that have D-rank congruent to m modulo t.

Conjecture 11.1.

$$(11.2) \quad N_2(0,5,5n+3) = N_2(1,5,5n+3) = N_2(2,5,5n+3) = N_2(3,5,5n+3) = N_2(4,5,5n+3) = \frac{1}{5} \phi_2(5n+3);$$

(11.3) $$cN_2(0,5,5n+3) = cN_2(1,5,5n+3) = cN_2(2,5,5n+3) = cN_2(3,5,5n+3) = cN_2(4,5,5n+3) = \frac{1}{5}\, c\phi_2(5n+3).$$

Equation (11.2) has been checked for $n = 0,1,2,3,$ and equation (11.3) for $n = 0$. These conjectures seem rather more difficult even than the Atkin and Swinnerton-Dyer theorem given above. In particular the elementary and elegant forms of the generating functions given in Section 2 of their paper have extremely complicated counterparts for our problem. (See note added in proof at end of paper.)

12. Partition Identities.

The Rogers-Ramanujan identities [4; Ch. 7] form the prototype for the results we shall be discussing in this section. These results are most famous as analytic identities:

(12.1) $$1 + \sum_{n=1}^{\infty} \frac{q^{n^2}}{(1-q)(1-q^2)\ldots(1-q^n)} = \prod_{n=0}^{\infty} \frac{1}{(1-q^{5n+1})(1-q^{5n+4})};$$

(12.2) $$1 + \sum_{n=1}^{\infty} \frac{q^{n^2}}{(1-q)(1-q^2)\ldots(1-q^n)} = \prod_{n=0}^{\infty} \frac{1}{(1-q^{5n+2})(1-q^{5n+3})}.$$

We shall state them as less familiar combinatorial results that suggest the subsequent developments in this section.

Definition 12.1. The i-th successive D-rank for the F-partition

$$\begin{pmatrix} a_1 a_2 \ldots a_r \\ b_1 b_2 \ldots b_r \end{pmatrix}$$

is $a_i - b_i$.

First Rogers-Ramanujan Identity. [4; pp. 155-156]. The number of ordinary partitions of n into parts congruent to 1 or 4 modulo 5 equals the number of partitions π of n which when written as Forbenius symbols have all successive D-ranks in the closed interval $[0,1]$.

Second Rogers-Ramanujan Identity. [4; pp. 155-156]. The number of ordinary partitions of n into parts congruent to 2 or 3 modulo 5 equals the number of partitions π of n which when written as Frobenius symbols have all successive D-ranks in the closed interval $[1,2]$.

These two somewhat unfamiliar statements of the Rogers-Ramanujan identities are special cases of an infinite family of results presented in [1] and [2] and extended by D. Bressoud [12] and W. Burge [13], [14].

Successive Ranks Theorem [2], [12]. The number of ordinary partitions of n into parts $\not\equiv 0, \pm a$ modulo K equals the number of partitions π of n which when written as Frobenius symbols have all successive D-ranks in $[-a+2, K-2-a]$.

The case $K = 5$, $a = 2$ is the first Rogers-Ramanujan identity; the case $K = 5$, $a = 1$ is the second.

There is, however, a different extension of the Rogers-Ramanujan identities due to B. Gordon [20], which has an elegant translation into F-partitions.

Theorem 12.1. Let $1 \leqq a \leqq k$ be integers, and let $A_{k,a}(n)$ denote the number of ordinary partitions of n into parts $\not\equiv 0, \pm a \pmod{2k+1}$. Let $\phi_{k,a}(n)$ denote the number of F-partitions with entries on each row repeated at most $k-1$ times, all successive D-ranks in $[0,1]$ and at most $a-1$ columns $\begin{matrix}\text{"0"}\\ 0\end{matrix}$. Then $A_{k,a}(n) = \phi_{k,a}(n)$ for all n.

Proof. Define $B_{k,a}(n)$ to be the number of ordinary partitions $b_1 + b_2 + \dots + b_r$ of n wherein $b_i - b_{i+1} \geqq 0$, $b_i - b_{i+k-1} \geqq 2$ and at most $a-1$ of the b's equal 1.

Then B. Gordon [20] has proved that $A_{k,a}(n) = B_{k,a}(n)$ for all n.

We now construct a bijection between the objects enumerated by $B_{k,a}(n)$ and those enumerated by $\phi_{k,a}(n)$. Namely for each b_i, the i-th part in a partition enumerated by $B_{k,a}(n)$, construct the $i^{\underline{th}}$ column

$$\begin{matrix}[b_i/2]\\ [(b_i-1)/2]\end{matrix}$$

in a partition enumerated by $\phi_{k,a}(n)$. Obviously the fact that at most $a-1$ of the b's are 1 corresponds to at most $a-1$ of the columns are $\begin{matrix}\text{"0"}\\ 0\end{matrix}$. The requirement $b_i - b_{i+k-1} \geqq 2$ clearly implies that there are at most $k-1$ repetitions of each integer on each row. Since $[x/2] + [(x-1)/2] + 1 = x$ for all integral x, we see that the same number n is being partitioned in each case. Clearly the mapping is reversible. Consequently $\phi_{k,a}(n) = B_{k,a}(n) = A_{k,a}(n)$ for all n. □

As an example of this result let us look at the case $k = 3$, $a = 2$, $n = 7$:

$B_{3,2}(7) = 6$		$\phi_{3,2}(7) = 6$
7	$\leftrightarrow$	$\begin{pmatrix} 3 \\ 3 \end{pmatrix}$
6 + 1	$\leftrightarrow$	$\begin{pmatrix} 30 \\ 20 \end{pmatrix}$
5 + 2	$\leftrightarrow$	$\begin{pmatrix} 21 \\ 20 \end{pmatrix}$
4 + 3	$\leftrightarrow$	$\begin{pmatrix} 21 \\ 11 \end{pmatrix}$
4 + 2 + 1	$\leftrightarrow$	$\begin{pmatrix} 210 \\ 100 \end{pmatrix}$
3 + 3 + 1	$\leftrightarrow$	$\begin{pmatrix} 110 \\ 110 \end{pmatrix}$

The natural question is: Are there other such results that would, for example, provide common generalizations of the Successive Ranks Theorem and Theorem 12.1? There do appear to be such results; in fact, Regime III of Baxter's solution to the Hard Hexagon Model [10] leads naturally to new theorems involving successive D-ranks. In trying to generalize these results, I found, using empiral data obtained through IBM's symbolic algebra language SCRATCHPAD, that F-partitions involving repetitions on rows arise naturally. It was at that point that a study of the fundamentals of F-partitions was undertaken and the results in this paper were obtained.

13. Asymptotics.

Kloosterman [26] in his study of the special quadratic forms considered in Section 9 utilized the Hardy-Littlewood circle method to obtain his results. We shall sketch in this section how the methods used there can be easily modified to obtain asymptotic formulae for $\phi_k(n)$ and $c\phi_k(n)$. Most likely this approach could be used to produce convergent asymptotic series for these functions in the same way that Rademacher [30] found convergent a asymptotic series for $p(n) = \phi_1(n) = c\phi_1(n)$.

Kloosterman [26] begins with the function

$$(13.1)\quad f_k(\nu|\tau) = \sum_{m_1,m_2,\ldots,m_k=-\infty}^{\infty} e^{\pi i\tau(m_1^2+\ldots+m_k^2) + 2\pi i\nu(m_1+\ldots+m_k)},$$

$$= \sum_{m,n=-\infty}^{\infty} r_k(n,m)\, e^{\pi i\tau n + 2\pi i\nu m},$$

where $r_k(n,m)$ is the number of simultaneous integral solutions of

$$(13.2)\quad m_1^2 + \ldots + m_k^2 = n, \quad m_1 + \ldots + m_k = m.$$

Immediately from Theorem 5.2, we see that by Cauchy's theorem [24; p. 76]

$$(13.3)\quad c\phi_k(n) = \frac{1}{2\pi i} \int_C \frac{dq}{q^{n+1}} \int_0^1 \frac{f_k(\nu|\frac{\tau}{2})}{(q)_\infty^k}\, d\nu.$$

$$= \frac{1}{2\pi i} \int_C (q)_\infty^{-k} q^{-n-1}\, dq \int_0^1 \vartheta(\nu\ \tfrac{\tau}{2})^k\, d\nu,$$

where C is a circle of radius < 1 centered on the origin, $q = e^{\pi i\tau}$ with $\operatorname{Im}\tau > 0$, and

$$(13.4)\quad \vartheta(\nu|\tau) = \sum_{n=-\infty}^{\infty} e^{\pi i\tau n^2 + 2\pi i n\nu}.$$

In a similar way we see from Theorem 5.1 that

$$(13.5)\quad \phi_k(n) = \frac{1}{2\pi i} \int_C (q)_\infty^{-k} q^{-n-1} dq$$

$$\times \int_0^1 \vartheta(\nu + \tfrac{1}{k+1} \mid \tfrac{\tau}{2})\vartheta(\nu + \tfrac{2}{k+1} \mid \tfrac{\tau}{2})\ldots\vartheta(\nu + \tfrac{k}{k+1} \mid \tfrac{\tau}{2})\, d\nu.$$

If one carries out the circle-method analysis for (13.3) noting that the primary contribution to the q-integral occurs in the neighborhood of $q = 1$, we find fairly directly that, for example,

$$(13.6)\quad c\phi_k(n) \sim \frac{1}{4\sqrt{3}n} e^{\pi\sqrt{\frac{2kn}{3}}}.$$

Clearly Hardy-Ramanujan-Rademacher series for $\phi_k(n)$ and $c\phi_k(n)$ can be derived from (13.3) and (13.5) by a full Farey dissection of the integrals in question.

14. F. Franklin's Method

It is worth pointing out that the proof of Euler's Pentagonal Number given by F. Franklin [17] extends easily to considerations of F-partitions.

Definition 14.1 Let $\Delta(n)$ denote the number of F-partitions of n

$$\begin{pmatrix} a_1 a_2 \ldots a_r \\ b_1 b_2 \ldots b_r \end{pmatrix}$$

wherein $a_i - a_{i+1} \geqq 2$, $1 \leqq b_i - b_{i+1} \leqq 2$, $0 \leqq b_r \leqq 1$, $a_r \geqq b_r$.

Theorem 14.1. $\Delta(n)$ is the number of ordinary partitions of n into distinct parts.

Proof. The correspondence between ordinary partitions and certain F-partitions given in Section 2 is precisely the bijection that establishes our theorem. The conditions in Definition 14.1 are precisely those required to make the bijection work. □

Example. $n = 9$, $\Delta(9) = 8$

Ordinary partitions of 9 with distinct parts	partitions enumerated by $\Delta(9)$
9	$\begin{pmatrix} 8 \\ 0 \end{pmatrix}$
8 + 1	$\begin{pmatrix} 7 \\ 1 \end{pmatrix}$
7 + 2	$\begin{pmatrix} 60 \\ 10 \end{pmatrix}$
6 + 3	$\begin{pmatrix} 51 \\ 10 \end{pmatrix}$
6 + 2 + 1	$\begin{pmatrix} 50 \\ 20 \end{pmatrix}$

$$5+4 \qquad \binom{42}{10}$$

$$5+3+1 \qquad \binom{41}{20}$$

$$4+3+2 \qquad \binom{31}{21}$$

We now translate F. Franklin's proof [17] of Euler's Pentagonal Number Theorem into a theorem on F-partitions.

Theorem 14.2. Let $\Delta_e(n)$ (resp. $\Delta_0(n)$) denote the number of F-partitions described in Definition 14.1 with the added condition that b_1 is even (resp. odd). Then

$$(14.1) \qquad \Delta_0(n) - \Delta_e(n) = \begin{cases} (-1)^j & \text{if } n = j(3j\pm1)/2 \\ 0 & \text{otherwise.} \end{cases}$$

Proof. Let π be an F-partition

$$\begin{pmatrix} a_1a_2\dots a_r \\ b_1b_2\dots b_r \end{pmatrix}$$

of the type described in Definition 14.1. Let $s(\pi)$ (the "slope" of π) be the least i such that $a_i - a_{i+1} > 2$. Let $b(\pi)$ (the "base" of π) be the least integer h such that $b_h - b_{h+1} = 2$.

If $b(\pi) \leqq s(\pi)$, associate π with

$$\pi' : \begin{pmatrix} (a_1+1)(a_2+1)\dots(a_{b(\pi)}+1)a_{b(\pi)+1}\dots a_r \\ (b_1-1)(b_2-1)\dots(b_{b(\pi)}-1)b_{b(\pi)+1}\dots b_r \end{pmatrix}$$

If $b(\pi) > s(\pi)$ associate π with

$$\pi'' : \begin{pmatrix} (a_1-1)(a_2-1)\dots(a_{s(\pi)}-1)a_{s(\pi)+1}\dots a_r \\ (b_1+1)(b_2+1)\dots(b_{s(\pi)}+1)b_{s(\pi)+1}\dots b_r \end{pmatrix}$$

Clearly if $b(\pi) \leqq s(\pi)$, then $s(\pi') = b(\pi)$, $b(\pi') > b(\pi)$; and if $b(\pi) > s(\pi)$ then $b(\pi'') = s(\pi)$, $s(\pi'') \geqq s(\pi)$. Now this association provides a bijection between the F-partitions enumerated by $\Delta_e(n)$ and those enumerated by $\Delta_0(n)$ except when the associated π'(or π'') happens to no longer be of the type enumerated by Definition 14.1. This occurs only when π is

$$\begin{pmatrix} (2j-2) \; \ldots\ldots \; 420 \\ (j-1)(j-2) \; \ldots \; 210 \end{pmatrix}$$

or

$$\begin{pmatrix} (2j-1)(2j-3) \; \ldots \; 531 \\ (j-1)(j-2) \; \ldots\ldots \; 210 \end{pmatrix}.$$

In these cases (and only these), the associated partition does not fulfill the conditions of Definition 14.1. In the first case $b_1 = j-1$ and $n = j + (j^2-j) + (j^2-j)/2 = j(3j-1)/2$, in the second case $b_1 - j-1$ and $n = j + j^2 + (j^2-j)/2 = j(3j+1)/2$. This proves the desired result. □

It is clear that one can extend the above argument of Franklin to a much wider class of F-partitions. Namely, any time there is a minimal difference prescribed for the top row and a maximal difference prescribed for the bottom row, one can define $s(\pi)$ as the index of the first nonoccurrence of the minimal top row difference, and $b(\pi)$ can denote the index of first maximal bottom row difference. Franklin's mappings then work as above, and one is left counting only those cases where the associated F-partitions fall outside the prescribed domain. Such results may prove useful in further studies of partition identities such as Theorem 12.1.

We also note that the Durfee square technique used by Sylvester [38; p. 33 of Coll. Works, Vol. 4] to prove

$$(14.1) \qquad \prod_{n=1}^{\infty} (1+zq^n) = 1 + \sum_{n=1}^{\infty} \frac{(1+zq)\ldots(1+zq^{n-1})}{(1-q)\ldots(1-q^n)} z^n q^{n(3n-1)/2}(1+zq^{2n}),$$

can easily be adapted to F-partitions.

15. Conclusion and Open Problems.

This study was originally undertaken to provide the foundation for F-partitions due to their natural appearance in extensions of the functions arising for Regime III of the Hard Hexagon Model [6] (see Section 12). As the foundations were explored, it was found that F-partitions come into intimate contact with Hecke modular forms (Section 7), the Lusztig-Macdonald-Wall conjectures (Section 8) and Kloosterman quadratic forms (Section 9) as well as the classical aspects of q-series, theta functions and partitions. Many questions naturally arise; the following is a partial list:

Problem 1. Provide purely combinatorial proofs of Theorems 5.1 and 5.2.

Problem 2. Provide purely combinatorial proofs of Corollaries 5.1 and 5.2.

Problem 3. The generalized Gaussian polynomials ${}_{rk}\begin{bmatrix} m+n \\ n \end{bmatrix}_q$ and ${}_{ck}\begin{bmatrix} m+n \\ n \end{bmatrix}_q$ are clearly (by Theorem 6.1) reciprocal polynomials of degree kmn. Show that they are unimodal, i.e. their coefficients are nondecreasing in the interval $[0, kmn/2]$. A good reference for recent work on unimodality theorems is [36].

Problem 4. Provide a purely combinatorial proof of Theorem 7.1.

Problem 5. Provide combinatorial proofs of the q-series identities: Theorems 8.1, 8.2, 8.3 and Corollaries 8.1, 8.2. Some nice combinatorial aspects of Corollary 8.2 are presented in [19] by Garsia and Remmel.

Problem 6. Provide analytic proofs of Corollaries 10.2 and 10.3 by giving explicit closed forms for $G_k(q)$ and $G_k(m,n;q)$. (cf. Atkin [8]).

Problem 7. Prove Conjecture 11.1, (cf. Atkin and Swinnerton-Dyer [9]).

Problem 8. Find Hardy-Ramanujan-Rademacher expansions for $\phi_k(n)$ and $c\phi_k(n)$. (cf. [4; Ch. 5]).

Problem 9. Find a sieve type proof of Theorem 12.1. (cf. [4; Ch. 9]).

Problem 10. Define integers a_n uniquely by

$$\Phi_4(q) = \prod_{n=1}^{\infty} (1-q^n)^{-a_n}.$$

Prove that $a_{10n-5} = 2$ for all n. All other a's seem to become large in absolute value for large subscript.

The above questions arise naturally from our work here. As further interactions occur between F-partitions and other topics in mathematics, many problems beyond these will present themselves.

16. Tables.

For reference we include values of both $\phi_k(n)$ and $c\phi_k(n)$ beyond those given incidentally in the preceding pages.

Table 1.

n	$\phi_1(n)$	$\phi_2(n)$	$\phi_3(n)$	$\phi_4(n)$	$\phi_5(n)$
1	1	1	1	1	1
2	2	3	3	3	3
3	3	5	6	6	6
4	5	9	11	12	12
5	7	14	18	20	21
6	11	24	31	35	37
7	15	35	49	56	60
8	22	55	78	92	99
9	30	81	119	142	156
10	42	120	180	221	244
11	56	171	267	330	371
12	77	248	394	496	561
13	101	345	567	724	829
14	135	486	813	1056	1218
15	176	669	1151	1512	1763
16	231	920	1616	2155	2532
17	297	1246	2244	3028	3594
18	385	1690	3099	4240	5068
19	490	2256	4240	5866	7075
20	627	3014	5769	8085	9819

Table 2.

n	$c\phi_1(n)$	$c\phi_2(n)$	$c\phi_3(n)$	$c\phi_4(n)$	$c\phi_5(n)$
1	1	4	9	16	25
2	2	9	27	68	150
3	3	20	82	256	675
4	5	42	207	777	2450
5	7	80	486	2160	7876

6	11	147	1055	5460	22825
7	15	260	2205	13056	61550
8	22	445	4374	29482	155925
9	30	744	8427	63952	375875
10	42	1215	15696	133456	867627
11	56	1944	28539	270080	1930775
12	77	3059	50630	531091	4159100
13	101	4740	88119	1019424	8708150
14	135	7239	150417	1913156	17771725
15	176	10920	252727	3520512	35447503
16	231	16286	418068	6360765	69243250
17	297	24028	682344	11305488	132718300
18	385	35110	1099343	19789160	249979625
19	490	50844	1750968	34159616	463348600
20	627	73010	2758185	58201535	846149380

References

1. G. E. Andrews, Sieves for theorems of Euler, Rogers and Ramanujan, from The Theory of Arithmetic Functions (A. A. Gioia and D. L. Goldsmith, eds.), Lecture Notes in Math., Springer, Berlin, No. 251, (1971), pp. 1-20.

2. G. E. Andrews, Sieves in the theory of partitions, Amer. J. Math., 94 (1972), 1214-1230.

3. G. E. Andrews, Applications of basic hypergeometric functions, S.I.A.M. Rev., 16(1974), 441-484.

4. G. E. Andrews, The Theory of Partitions, Encyclopedia of Math. and Its Applications, Vol. 2, Addison-Wesley, Reading, 1976.

5. G. E. Andrews, Partitions, q-series and the Lusztig-Macdonald-Wall conjectures, Invent. Math., 41(1977), 91-102.

6. G. E. Andrews, The hard-hexagon model and Rogers-Ramanujan type identities, Proc. Nat. Acad. Sci. U.S.A., 78(1981), 5290-5292.

7. G. E. Andrews, Hecke modular forms and the Kac-Peterson identities, Trans. Amer. Math. Soc. (to appear).

8. A. O. L. Atkin, Ramanujan congruences for $p_{-k}(n)$, Canadian J. Math., 20(1968, 67-78.

9. A. O. L. Atkin and H. P. F. Swinnerton-Dyer, Some properties of partitions, Proc. London Math. Soc. (3), 4(1954), 84-106.

10. R. J. Baxter, Rogers-Ramanujan identities in the hard-hexagon model, J. Stat. Phys., 26(1981), 427-452.

11. F. van der Blij, On the theory of simultaneous linear and quadratic representation III, Indag. Math., 9(1947), 129-135.

12. D. Bressoud, Extension of the partition sieve, J. Number Th., 12(1980), 87-100.

13. W. Burge, A correspondence between partitions related to generalizations of the Rogers-Ramanujan identities, Discrete Math., 34(1981), 9-15.

14. W. Burge, A three way correspondence between partitions, European J. of Combinatorics, (to appear).

15. D. I. A. Cohen, Some sieves for partition theory, J. Comb. Th. (A), 27(1979), 325-332.

16. F. J. Dyson, Some guesses in the theory of partitions, Eureka (Cambridge) 8(1944), 10-15.

17. F. Franklin, Sur le développement du produit infini $(1-x)(1-x^2)(1-x^3)\ldots$, Comptes Rendus, 82(1881), 448-450.

18. G. Frobenius, Über die Charaktere der symmetrischen Gruppe, Sitzber. Preuss. Akad. Berlin, 1900, 516-534.

19. A. Garsia and J. Remmel, A combinatorial view of Andrews' proof of the L-M-W conjectures, European J. Combinatorics, (to appear).

20. B. Gordon, A combinatorial generalization of the Rogers-Ramanujan identities, Amer. J. Math., 83(1961), 393-399.

21. B. Gordon, Two new representations of the partition function, Proc. Amer. Math. Soc., 13(1962), 869-873.

22. B. Gordon and L. Houten, Notes on plane partitions II, J. Comb. Th., 4(1968), 81-99.

23. H. Gupta, Selected Topics in Number Theory, Abacus Press, Tunbridge Wells, 1980.

24. G. H. Hardy and S. Ramanujan, Asymptotic formulae in combinatory analysis, Proc. London Math. Soc. (2), 17(1918), 75-115.

25. C. G. J. Jacobi, Fundamenta Nova Theoriae Funktionum Ellipticarum, Regiomonti fratrum Borntr̈ager, 1829 (reprinted in Gesammelte Werke, Vol. 1, pp. 49-239; Reimer, Berlin, 1881).

26. H. D. Kloosterman, Simultane Darstellung zweier ganzen Zahlen als einer Summe von ganzen Zahlen und deren Quadratsumme, Math. Ann. 118(1942), 319-364.

27. D. E. Littlewood, The Theory of Group Characters and Matrix Representations of Groups, 2nd ed., Oxford University Press, London and New York, 1950.

28. G. A. Lomadze, On the summation of a singular series I, Akad. Nauk Gruzin. SSR. Trudy Tbliss. Mat. Inst. Razmadze 19(1953), 61-77.

29. G. Lusztig, Irreducible representation of finite classical groups, Invent. Math., 43(1977), 125-177.

30. H. Rademacher, On the partition function $p(n)$, Proc. London Math. Soc., (2), 43(1937), 241-254.

31. S. Ramanujan, Collected Papers of S. Ramanujan, Cambridge University Press, London and New York, 1927 (reprinted by Chelsea, New York).

32. G. de B. Robinson, Representation Theory of the Symmetric Group, Univ. of Toronto Press, Toronto, 1961.

33. L. J. Slater, Further identities of the Rogers-Ramanujan type, Proc. London Math. Soc. (2), 54(1952), 147-167.

34. L. J. Slater, Generalized Hypergeometric Functions, Cambridge University Press, London and New York, 1966.

35. H. J. S. Smith, Collected Math. Papers, Vol. 1, Oxford University Press, London, 1894 (reprinted Chelsea, New York, 1965).

36. R. P. Stanley, Unimodal sequences arising from Lie algebras, from Combinatorics, Representation Theory and Statistical Methods in Groups, Narayana, Mathsen and Williams eds., Lecture Notes in Pure and Applied Math., Vol. 57, Dekker, New York, 1980.

37. C. Sudler, Two enumerative proofs of an identity of Jacobi, Proc. Edinburgh Math. Soc., 15(1966), 67-71.

38. J. J. Sylvester, A constructive theory of partitions in three acts, an interact and an exodion, Amer. J. Math., 5(1882), 251-330 (or pp. 1-83 of the Coll. Math. Papers of J. J. Sylvester, Vol. 4, Cambridge Univ. Press, London and New York, 1912; reprinted by Chelsea, New York, 1974).

39. A. Z. Valfiš, The additive theory of numbers XI, Akad. Nauk Gruzin. SSR. Trudy Tbliss. Mat. Inst. Razmadze, 19(1953), 33-59.

40. E. M. Wright, An enumerative proof of an identity of Jacobi, J. London Math. Soc., 40(1965), 55-57.

41. E. M. Wright, Stacks, Quart. J. Math. (Oxford), 19(1968), 313-320.

42. E. M. Wright, Stacks II, Quart. J. Math. (Oxford), 22(1971), 107-116.

43. E. M. Wright, Stacks III, Quart. J. Math (Oxford), 23(1972), 153-158.

44. J. Zolnowsky, A direct combinatorial proof of the Jacobi identity, J. Comb. Th. (A), 27(1979), 333-341.

(Note added in proof: A lengthy computer calculation by F. Garvan shows that the conjecture (11.2) is false for $n = 4$ and 5).

THE PENNSYLVANIA STATE UNIVERSITY

UNIVERSITY PARK, PENNSYLVANIA 16802

General instructions to authors for
PREPARING REPRODUCTION COPY FOR MEMOIRS

For more detailed instructions send for AMS booklet, "A Guide for Authors of Memoirs." Write to Editorial Offices, American Mathematical Society, P. O. Box 6248, Providence, R. I. 02940.

MEMOIRS are printed by photo-offset from camera copy fully prepared by the author. This means that, except for a reduction in size of 20 to 30%, the finished book will look exactly like the copy submitted. Thus the author will want to use a good quality typewriter with a new, medium-inked black ribbon, and submit clean copy on the appropriate model paper.

Model Paper, provided at no cost by the AMS, is paper marked with blue lines that confine the copy to the appropriate size. Author should specify, when ordering, whether typewriter to be used has PICA-size (10 characters to the inch) or ELITE-size type (12 characters to the inch).

Line Spacing – For best appearance, and economy, a typewriter equipped with a half-space ratchet – 12 notches to the inch – should be used. (This may be purchased and attached at small cost.) Three notches make the desired spacing, which is equivalent to 1-1/2 ordinary single spaces. Where copy has a great many subscripts and superscripts, however, double spacing should be used.

Special Characters may be filled in carefully freehand, using dense black ink, or INSTANT ("rub-on") LETTERING may be used. AMS has a sheet of several hundred most-used symbols and letters which may be purchased for $5.

Diagrams may be drawn in black ink either directly on the model sheet, or on a separate sheet and pasted with rubber cement into spaces left for them in the text. Ballpoint pen is *not* acceptable.

Page Headings (Running Heads) should be centered, in CAPITAL LETTERS (preferably), at the top of the page – just above the blue line and touching it.

LEFT-hand, EVEN-numbered pages should be headed with the AUTHOR'S NAME;

RIGHT-hand, ODD-numbered pages should be headed with the TITLE of the paper (in shortened form if necessary).

Exceptions: PAGE 1 and any other page that carries a display title require NO RUNNING HEADS.

Page Numbers should be at the top of the page, on the same line with the running heads.

LEFT-hand, EVEN numbers – flush with left margin;

RIGHT-hand, ODD numbers – flush with right margin.

Exceptions: PAGE 1 and any other page that carries a display title should have page number, centered below the text, on blue line provided.

FRONT MATTER PAGES should be numbered with Roman numerals (lower case), positioned below text in same manner as described above.

MEMOIRS FORMAT

**It is suggested that the material be arranged in pages as indicated below.
Note: Starred items (*) are requirements of publication.**

Front Matter (first pages in book, preceding main body of text).

Page i – *Title, *Author's name.

Page iii – Table of contents.

Page iv – *Abstract (at least 1 sentence and at most 300 words).

*1980 Mathematics Subject Classifications represent the primary and secondary subjects of the paper. For the classification scheme, see Annual Subject Indexes of MATHEMATICAL REVIEWS beginning in December 1978.

Key words and phrases, if desired. (A list which covers the content of the paper adequately enough to be useful for an information retrieval system.)

Page v, etc. – Preface, introduction, or any other matter not belonging in body of text.

Page 1 – Chapter Title (dropped 1 inch from top line, and centered).

Beginning of Text.

Footnotes: *Received by the editor date.
Support information – grants, credits, etc.

Last Page (at bottom) – Author's affiliation.

ABCDEFGHIJ–AMS–8987654